SpringerBriefs in Medical Earth Sciences

More information about this series at http://www.springer.com/series/13202

Gouri Sankar Bhunia • Pravat Kumar Shit

Editors

Spatial Mapping and Modelling for Kala-azar Disease

Gouri Sankar Bhunia
TPF Gentisa Euroestudios S L
Noida, Haryana, India

Pravat Kumar Shit
PG Department of Geography
Raja N. L. Khan Women's College
(Autonomous)
Midnapore, West Bengal, India

ISSN 2523-3610   ISSN 2523-3629   (electronic)
SpringerBriefs in Medical Earth Sciences
ISBN 978-3-030-41226-5   ISBN 978-3-030-41227-2   (eBook)
https://doi.org/10.1007/978-3-030-41227-2

This Springer imprint is published by the registered company Springer Nature Switzerland AG
The registered company address is: Gewerbestrasse 11, 6330 Cham, Switzerland

Dedicated to our family members

Preface

Kala-azar disease is geographically and temporally limited by variations in environmental variables such as temperature, humidity, rainfall, vegetation, and land use pattern. Due to the absence of any suitable "epidemic prediction tool," it is very difficult to forewarn epidemic outbreaks or to make any strategic plan and combat this menace. Customary GIS tools and the open-source GIS software have extended their presentations to both operate core functionality and inflate upon these utilities through software extensions. This book includes a practical coverage of the use of open-source software solutions in public health problems, particularly in vector-borne disease control programs.

This book titled *Spatial Mapping and Modelling for Kala-Azar Disease* gives an overview and discusses approaches and case studies to exemplify numerous methods and strategies in addressing kala-azar. It also includes a practical coverage of the use of spatial analysis techniques for infectious disease data through open-source software solutions. In this book, environmental factors (relief characters, climatology, ecology, vegetation, water bodies, etc.) and socioeconomic issues (housing type and pattern, education level, economic status, income level, working status, etc.) are also investigated at small administrative units (village) that address kala-azar disease control. The first part of the book highlights the basic concepts of kala-azar and the role of geoinformatics in kala-azar disease. The next part of the book covers various microgeographical factors creating suitable conditions for vector propagation and disease transmission. Another part of the book contains spatio-temporal modelling; geospatial data mining; various statistical and mathematical applications; accuracy and uncertainty of geoscientific models; applications in environmental, ecological, and biological modelling; and analysis in public health research. This book also presents the use of real case studies and hands-on exercises of public health data through open-source GIS. Therefore, this book aims to provide an example of spatial mapping and modelling by considering practical presentations and subjects that may be of use in public health programs.

Noida, Haryana, India
Midnapore, West Bengal, India

Gouri Sankar Bhunia
Pravat Kumar Shit

Acknowledgments

We are very much thankful to our teachers Dr. Nandini Chatterjee, Dr. Dilip Kr. Pal, Dr. Shreekant Kesari, Dr. Pradeep Das, Mr. Ajoy Mandal, Dr. Sunando Bandyopadhyay, Dr. Ramkrishna Maiti, Dr. Ashis Kumar Paul, Dr. Laxmi Narayan Satpati, Dr. Joytishankar Bandyopadhyay, and Dr. Ratan Kumar Samanta for their many experiences, suggestions, encouragement, and immense support throughout the work.

We are also thankful to Mr. Rakesh Mandal and Tarang Sukhatme for their help and cooperation. We also acknowledge the contribution of the Postgraduate Department of Geography, Raja N.L. Khan Women's College (Autonomous), for providing logistic support and infrastructure facilities. Dr. Pravat Kumar Shit would like to thank Dr. Jayasree Laha, Principal, Raja N.L. Khan Women's College (Autonomous), Midnapore, for her administrative support to carry on this project.

We would also like to thank Ranita and Debjani, whose love, encouragement, and support kept us motivated up to the final shape of the book. Finally, the book has taken a number of years in its making, and we, therefore, want to thank our families and friends for their continued support.

Contents

About the Authors

Gouri Sankar Bhunia has received his Ph.D. in Geography (Health GIS) from the University of Calcutta, India, in 2015. His Ph.D. dissertation work focused on environmental control measures of infectious disease (Visceral leishmaniasis or kala-azar) using geospatial technology. His research interests include kala-azar disease transmission modelling, environmental modelling, risk assessment, urban geography, groundwater resource management research, soil science research data mining, and information retrieval using geospatial technology. Dr. Bhunia is Associate Editor and is on the editorial boards of three international journals in health GIS and geosciences. He is a *GIS Expert* in Jammu Smart City Project, Jammu, and is a *Visiting Faculty* in Nalini Prabha Dev Prasad Roy College of Bilaspur and Seacom Skill University of West Bengal. He also worked as a *Manager* in GIS Division of Aarvee Associates, Hyderabad, and *Resource Scientist* in Bihar Remote Sensing Application Centre, Patna. He is Recipient of the *Senior Research Fellow (SRF)* from the Rajendra Memorial Research Institute of Medical Sciences (ICMR, India) and contributed to multiple research programs on kala-azar disease transmission modelling.

Pravat Kumar Shit has received his Ph.D. in Geography in 2013 and M.Sc. in Geography and Environment Management in 2005, both from Vidyasagar University, India, and PG Diploma in Remote Sensing and GIS from Sambalpur University in 2015. He is Assistant Professor in the Department of Geography, Raja N. L. Khan Women's College (Autonomous), Gope Palace, Midnapore, West Bengal, India. His main fields of research are soil erosion spatial modelling, badland geomorphology, gully morphology, and water resources and natural resources mapping and modelling. He has published more than 50 international and national research articles in various renowned journals and 3 books for Springer Publishing. His research work has been funded by the University Grants Commission (UGC), India, and the Department of Higher Education Science and Technology and Biotechnology, Government of West Bengal. Dr. Shit is Associate Editor and is on the editorial boards of three international journals in geography and earth environment sciences.

Chapter 1
Introduction of Visceral Leishmaniasis (Kala-azar)

Abstract This chapter begins with a summary of leishmaniasis in parallel with special emphasis on kala-azar. Kala-azar or visceral leishmaniasis (VL) is a slow progressing indigenous disease caused by a protozoan parasite (*Leishmania donovani*, *Leishmania infantum*, and *Leishmania archibaldi*), with a mortality rate 75–95%. The parasite primarily infects the reticuloendothelial system and may be found in abundance in the bone marrow, spleen, and liver. Out of the 88 VL-affected countries, 72 countries are least developed countries, and 90% of kala-azar cases are recorded from India, Bangladesh, Nepal, and Sudan. Approximately 600 *Leishmania* species have been identified in the Old World and New World. However, the prevalence of *P. argentipes* is observed throughout the year with two annual peak density. In India, VL is purely anthroponosis. A multifaceted aspect has seen the reemergence and transmission of kala-azar throughout the world.

Keywords Kala-azar · Visceral leishmaniasis · *Leishmania* · *Phlebotomus argentipes* · New World and Old World

1.1 Introduction

Currently, vector-borne diseases are among some of the major microbial causes of morbidity and mortality. Leishmaniasis is one of the most important vector-borne diseases of humans, which can be caused by many species of *Leishmania*, a protozoan parasite of the family Trypanosomatidae (order Kinetoplastida) (Williams 1993). Leishmaniasis is one of the neglected tropical diseases becoming a challenge for medical science in about 98 countries with a mortality rate of over 50,000 per year (GBD 2013). Between 12 and 15 million people are infected, and 350 million people are at risk. 1.5–2 million new cases occur each year (Torres-Guerrero et al. 2017), out of which 1–1.5 million cases are of cutaneous leishmaniasis and 500,000 cases of visceral leishmaniasis.

© The Author(s), under exclusive license to Springer Nature Switzerland AG 2020   1
G. S. Bhunia, P. K. Shit, *Spatial Mapping and Modelling for Kala-azar Disease*,
SpringerBriefs in Medical Earth Sciences, https://doi.org/10.1007/978-3-030-41227-2_1

The classification of *Leishmania* is complex and, in some cases, controversial; more than one species name may be used for an organism, and some names may eventually be invalidated (Spickler et al. 2010). The World Health Organization (WHO) has acknowledged leishmaniasis as a category I disease, and the World Health Assembly (WHA) 43.18 resolution identifies leishmaniasis as a major public health concern (World Health Organization 2004). Types of leishmaniasis include:

- *Visceral leishmaniasis(VL):* Usually affects the spleen, liver, or other lymphoid tissues and, if left untreated, is fatal; part of effectively treated VL cases may follow in maculopapular or nodular rashes (Fig. 1.1a)
- *Cutaneous leishmaniasis:* Skin infection, causing lesions and ulcers in one area of the body (Fig. 1.1b)
- *Mucocutaneous leishmaniasis:* Skin infection, causing lesions, plus lesions in the nose and throat (Fig. 1.1c)
- *Diffuse cutaneous leishmaniasis:* Skin infection causing lesions in multiple sites on the body

Visceral leishmaniasis (VL), locally known as *kala-azar* (KA) in India, continues to be an economic burden and a great threat globally and almost impossible to eradicate since the last few decades. VL can be classified as anthroponotic visceral leishmaniasis (AVL) and zoonotic visceral leishmaniasis (ZVL). AVL is transmitted between humans via vector carriers, primarily caused by *Leishmania donovani*, whereas ZVL is transmitted between humans and other mammals through *L. donovani*, *L. infantum*, and *L. archibaldi*. AVL is mainly observed in East Africa (Somalia), northeastern Africa (Sudan), and the Middle East (particularly Yemen and Saudi Arabia). ZVL is concentrated in East Africa, Brazil, Mediterranean Basin, and South Asia (Quinnell and Courtenay 2009). It is the world's second-deadliest parasitic disease after malaria, with 400,000 cases and 40,000 deaths occurring annually (Alvar et al. 2012). Kala-azar is caused by *Leishmania* protozoa that are spread through the bite of female sandflies of the family Psychodidae (subfamily – Phlebotomine) (WHO 2000a, b). VL is caused by *Leishmania donovani* in

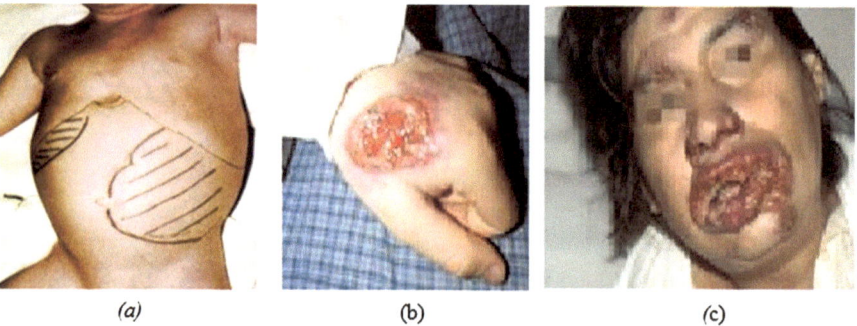

(a)   *(b)*   *(c)*

Fig. 1.1 Various forms of leishmaniasis. (**a**) Visceral Leishmaniasis. (**b**) Cutaneous Leishmaniasis. (**c**) Mucocutaneous Leishmaniasis. (*Source*: http://www.who.int/campaigns/world-health-day/2014/photos/leishmaniasis/en/)

Table 1.1 Global scenario of leishmaniasis

Disease form	Geographical distribution
Cutaneous leishmaniasis (CL)	90% of cases in Afghanistan, Algeria, Brazil, the Islamic Republic of Iran, Peru, Saudi Arabia, Sudan, and the Syrian Arab Republic
Mucocutaneous leishmaniasis (MCL)	90% of cases in Brazil, Peru, and the Plurinational State of Bolivia
Visceral leishmaniasis (VL) or Kala-azar (KA)	90% of cases in Bangladesh, Brazil, Ethiopia, India, Nepal, and Sudan

India and eastern Africa and by *Leishmania infantum/Leishmania chagasi* in the Mediterranean Basin, western Africa, and Latin America (WHO 2000a). 90% of the VL-affected cases are found in Bangladesh, India, Nepal, Sudan, and Brazil (Table 1.1). More than 60% of VL cases in the world originated in South Asia. Traditionally, the disease exists in rural areas, but it does not recur in one area; it keeps occurring in new environments. The disease is geographically and temporally limited by variations in environmental variables such as temperature, humidity, rainfall, vegetation, and land use pattern. Due to the absence of any suitable *epidemic prediction tool*, it is very difficult to forewarn epidemic outbreaks or to make any strategic plan and combat this menace.

VL is primarily caused by *Leishmania donovani* and *L. infantum/L. chagasi*. *L. donovani* is anthroponotic and *L. infantum* is zoonotic (Rijal et al. 2010). *L. donovani* causes VL in South Asia and Africa, whereas *L infantum* is mainly distributed in the Middle East, Mediterranean Basin, Latin America, and parts of Asia (Table 1.2). *L. infantum* in the "Old World" (Eastern Hemisphere) and *L. chagasi* in the "New World" (Western Hemisphere) – and these two organisms were thought to be different species.

VL is transmitted by the bite of tiny and seemingly innocuous female *phlebotomine* sandfly; the parasite comes into macrophages, where it proliferates and establishes the infection (Fig. 1.1). The sandflies inject the infective stage, promastigotes, during blood meals (1). Promastigotes that reach the puncture wound

Table 1.2 Main species of VL with correspondent clinical form

Species	Vector	Region	Host
Leishmania infantum	*P. perniciosus, P. ariasi*	Southern Europe, Middle and Central Asia, northwest Africa, Mediterranean Basin	Human, dogs, sylvatic canids
Leishmania donovani	*P. argentipes, P. orientalis, P. martini*	Ethiopia, Sudan, Kenya, India, China, Bangladesh, Burma	Human anthroponosis, Sudan rodents, canines
Leishmania infantum	*L. longipalpis, L. cruzi, L. cvansi*	South, Central, and North America	Human, sylvatic canids and fields, opossums, dogs
Leishmania siamensis	*P. sergentomiya*	Europe, Asia, and North America	Human, horse, cows

are phagocytized by macrophages (2) and transform into amastigotes (3). Amastigotes multiply in infected cells and affect various tissues, depending in part on the *Leishmania* species (4). This originates the clinical manifestations of leishmaniasis. Sand flies become infected during blood meals on an infected host when they ingest macrophages infected with amastigotes (5 and 6). In the sandfly's midgut, the parasites differentiate into promastigotes (7), which multiply and migrate to the proboscis (8) (Fig. 1.2); *Illustration*: **CDC/Alexander J da Silva/ Melanie Moser**).

1.2   Discovery of Visceral Leishmaniasis (Kala-azar)

Before 1820s, the history of kala-azar was mystery. In 1563, a Portuguese physician and botanist, *Garcia da Orta*, first described about the fever in Goa. *Sir Harold Scott* recommended that Orta could be termed as kala-azar (Scott 1939). In 1768, 200 years later, the naval surgeon James Lind designated the diseases of hot temperatures but did not identify kala-azar. Still it is unknown the commencement of kala-azar. By 1794, *Richard Shannon* had mentioned that enlargement of the spleen is linked with swelling of the liver, asserting that it was normally called *ague cake*.

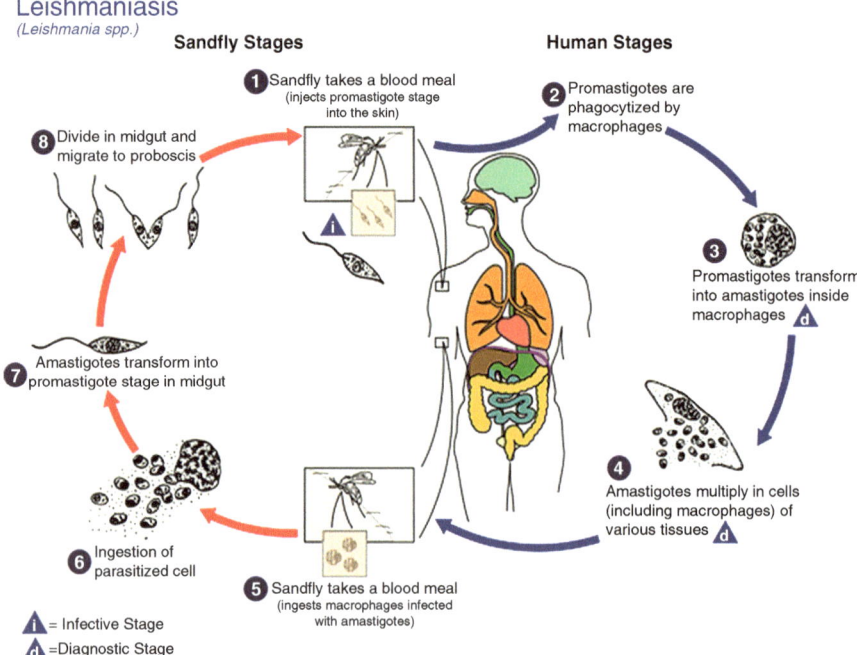

Fig. 1.2 The life cycle of *Leishmania* spp., the causal agents of leishmaniasis. (Source: Hailu et al. 2005)

Before 1820, the disease was named *Jessore fever* as it originated in Jessore district of India (now in Bangladesh). *Dr. J Elliot*, the civil surgeon of Burdwan, drew the first epidemic in 1824–1825 in a large village called Mahmedopore (Elliot 1863). *Upendra Nath Brahmachari* of Bengal Medical Service concluded from both clinical and statistical evidences that Burdwan fever was of malarial origin (Brahmachari 1928). From the beginning of 1860, the Government of Calcutta provided several reports of an epidemic fever, which Rogers later considered to be kala-azar (Rogers 1897). The epidemic sustained into the early years of the 1870s, and the Sanitary Commissioner of Bengal, Surgeon-Major *Charles J. Jackson*, added a prolonged compilation of reports by local medical officers to his annual report of 1873 (Bengal, Sanitary Commissioner report 1873).

Lieutenant General Sir William Boog Leishman was a Scottish pathologist and British Army medical officer. He was a distinguished bacteriologist and pathologist who gave his name to a disease, "leishmaniasis," the cause of which he discovered. He served in India, where he did a research on enteric fever and kala-azar. In 1901, while examining pathologic specimens of a spleen from a patient who had died of kala-azar (now called visceral leishmaniasis), he observed oval bodies and published his account of them in 1903 (Source: https://en.wikipedia.org/wiki/William_Boog_Leishman).

In 1869, further reports of an epidemic fever arrived in the Bengal Medical Department, but these were from a totally different area. Subsequently, the Dinajpur and Rangpur district, contiguous to Garo hills in Assam, were specified to have been facing an epidemic fever analogous in character. Nearly, a severe epidemic was observed to be raging in the Garo hills, the adjacent point in Assam territory. In 1885, Cunningham first stated that *Leishmania* organism was not a bacterium. Thirteen years later, Peter Borovsky (a Russian military surgeon) found out the protozoan which was also confirmed by Wright in 1903.

By 1891, the disease had moved eastward and definitely recognized itself in Nowgong district. In 1898, *Sir Ronald Ross* was hired by the government to conduct an examination about the nature of the disease, known as *kala-dukh* (black pain in Hindustani) or *kala-jwar* (black fever in Bengali) in Darjeeling and Purnea in Upper Bengal (India). The disease was tenacious, and the mortality rate declined by 31.5% between 1891 and 1901 (Rogers 1950). In 1903, *William Boog Leishman*, an English military surgeon, and *Charles Donovan*, an Irish medical officer, autonomously stated the existence of an unknown organism, in the smear preparation from the spleen pulp of persons who expired from this endemic fever. In the meantime, the terrible epidemic has made *kala-azar* a word of terror in Assam. The Nowgong epidemic has ended in 1901, but the disease spread gradually in an easterly direction up to the Brahmaputra valley and, by 1910, was well-established in Golaghat subdivision. Seven years later, the disease had stretched the Sibsagar subdivision. During the period between 1918 and 1919, McCombie Young estimated more than 2,00,000 deaths due to kala-azar in Assam. Later epidemic of kala-azar occurred in Assam in 1925, 1944, and 1963.

Charles Donovan was an Irish medical officer in the Indian Medical Service. He also recognized oval bodies in other kala-azar patients as the protozoan that causes kala-azar, *Leishmania donovani* (*Source*: https://en.wikipedia. org/-wiki/Charles_Donovan).

The epidemic began in Bihar, Assam, West Bengal, and Tamil Nadu in 1930, 1940, 1943, and 1947, respectively. In Bihar, numerous epidemics were known in 1882–1885, 1917, 1933, and 1939. The current epidemic is noted in the early 1970s. In 1977, a sample survey estimated 70,000 cases and 4500 deaths in the four districts of North Bihar, i.e., Vaishali, Muzaffarpur, Sitamarhi, and Samastipur. Currently, the disease is also occurring in Uttar Pradesh. Formerly, the disease was limited to the northwest, but later it became prevalent from Madras to Tuticorin (Tamil Nadu) and south of Andhra Pradesh. Few foci have also been evidenced from Punjab, Rajasthan, and Kashmir. In Gujarat, intermittent cases of VL or kala-azar have been described.

Fig. 1.3 Sand fly (vector of kala-azar) also known as *P. argentipes*. (Source: Bhunia 2014)

1.3   Incrimination of Sandfly (Phlebotomus argentipes)

There are 600 species of sandflies divided into *Phlebotomus* and *Sergentomyia* in the Old World and *Lutzomyia*, *Brumptomyia*, and *Warileya* in the New World (Ready 2013; Maroli et al. 2012). Mackie (1914) first suspected the possible association of the genus *Phlebotomus* with the transmission of kala-azar. Awati (1922) in a list of biting insects found in dwelling houses in Assam mentioned only one species of *Phlebotomus*, viz., *P. perturbans*. As mentioned by Shortt and Swaminath (1928), there is a little qualm that Mackie was not aware of its identity when he was working on kala-azar (KA) in Assam, when he dissected *Phlebotomus argentipes*. In 1922, Swaminath, working in the KA project of the Indian Research Fund Association in Assam, under the direction of Major H. E. Shortt, again dissected *P. argentipes* without positive identification. In 1924, Shortt classified suspected transmitters of KA into groups as "probable" and "possible," the genus *Phlebotomus* being placed in the former group. In 1924, Knowless et al. first observed that about 25% of the sandflies, with heavy infection of flagellates, fed on KA patients. In March 1925, Christophers, Shortt, and Barraud gave the first description of the flagellates encountered in the infected flies and of their location in the gut of the insect, and they also considered a series of controls with laboratory breed flies and observed that the flagellates seen in *P. argentipes*, after feeding on KA cases, really were *L. donovani*. In July 1926, the first specimen of *P. argentipes* caught in nature and showing an infection with *L. donovani* was obtained by the Commission (Fig. 1.3).

1.3.1   *Etiopathogenesis of* Phlebotomus argentipes

The vector of various species and subspecies of the protozoa of the *Leishmania* genus are dipterans of the genus *Lutzomyia* in the New World and *Phlebotomus* in the Old World belonging to the subfamily Phlebotomine. The taxonomy of the vector is as follows:

Kingdom	:	Animal
Phylum	:	Arthropoda
Subphylum	:	Euarthropoda
Superclass	:	Antenata
Class	:	Insecta/Hexapoda
Order	:	Diptera
Suborder	:	Nematocera
Family	:	Psychodidae
Subfamily	:	Phlebotomine
Genus	:	Old World: *Phlebotomus/Sergentomyia*
		New World: *Lutzomyia*

The subfamily Phlebotomine comprises six genera, viz., *Phlebotomus*, *Sergentomyia*, *Brumptomyia*, *Warileya*, and *Psychodopygus*. Out of these, only two genera, viz., *Phlebotomus* and *Sergentomyia*, have been recorded from the Indian subcontinent. The adult sandflies are small, fuzzy, delicately proportioned, about 1.5–3.5 mm in length, and about one-fourth of the size of a mosquito (Fig. 1.4a). They are light yellow to grayish brown in color with large conspicuous dark eyes. The head, thorax, and abdomen are densely covered with long hairs. All female sandflies suck blood for survival and development of eggs (Short et al. 1927). They feed on a variety of warm- and cold-blooded animals like birds, mammals, and reptiles (Sanyal et al. 1979a). The life cycle (Fig. 1.4b) is characterized by complete metamorphosis and comprises egg, larva, pupae, and adult (Sanyal et al. 1979b). After taking blood, female sandflies rest for sometime in dark and moist places for the maturation of ova. The preoviposition period varies in different species.

The total number of eggs laid per female *P. argentipes* ranges from 5 to 68. The larva is creamy white in color and possesses several hairs on its body. Duration of larval stages is directly related to temperature, humidity, and availability of food. The total larval period of *P. argentipes* is recorded to be 11–29 (mean: 15.15) days. The pupae are elongated and comma-shaped in appearance and represent a nonfeeding stage. The pupal stage varies from 6 to 10 (mean: 7.65) days for *P. argentipes* under optimum laboratory conditions. The sexes can be distinguished at the pupal stage. In males, the genitalia extend posterior as a build process, whereas in females, the tip of the abdomen of pupae is pointed without any extension. Life cycle from egg to emergence of adult *P. argentipes* was found to be 20–36 days (mean: 26.75).

1.3.2   Seasonality and Nocturnal Landing/Biting Activity of Phlebotomus argentipes

The prevalence of *P. argentipes* is observed throughout the year. Two annual density peaks are found around premonsoon (June–July) and postmonsoon (August–September) seasons (Picado et al., 2010a). *P. argentipes* is very less observed during the winter season (December–January), while occasional specimens were observed during the winter. Dinesh et al. (2001) reported peak landing and biting hour for *P. argentipes* on humans and bovines at 23.00–24.00 hrs. The female/male ratio on human bait was 1:8, and in cattle it was 1:13.

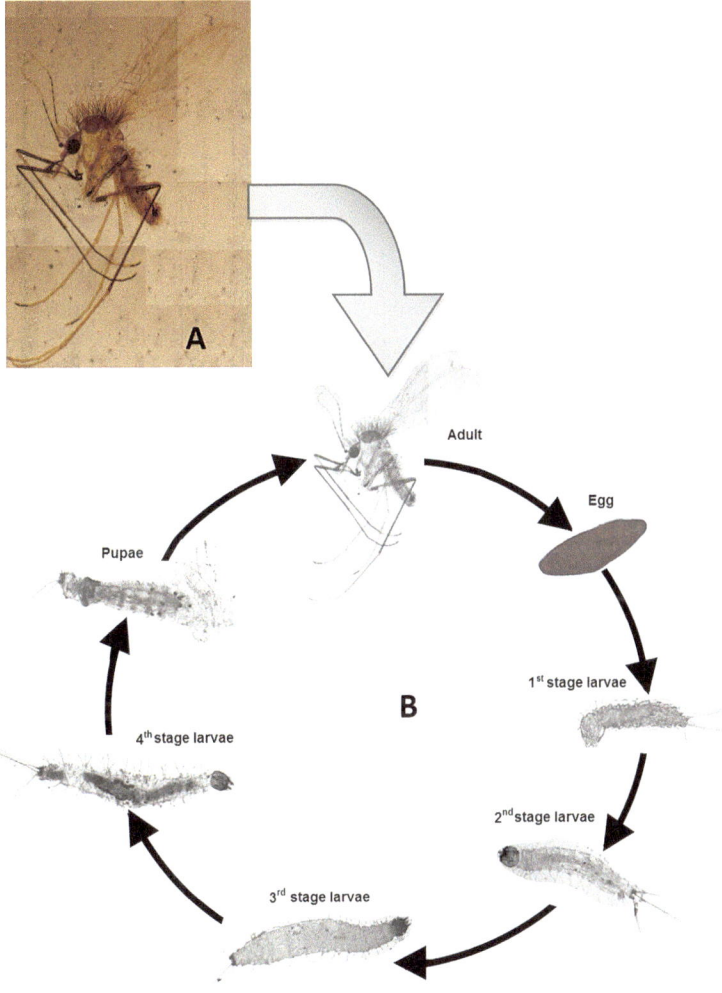

Fig. 1.4 (**a**) *Phlebotomus argentipes*. (**b**) Life cycle of sandfly (*Phlebotomus argentipes*) (Bhunia 2014)

1.4 Transmission of Visceral Leishmaniasis and Behavior of Parasite

Visceral leishmaniasis (VL) has developed an epidemic cycle taking place almost regularly every 15–20 years (Alvar et al. 2006). VL infection is spread from humans to other mammals by the bite of an infected female sandfly vector (Ready 2013). The infection can rarely be transmitted by other means such as blood transfusion (Cohen et al. 1991), needle sharing (Cruz et al. 2002), or from mother to child during pregnancy (Meinecke et al. 1999). The adult female sandfly is a bloodsucker, typically suckling at night on sleeping prey. Female sandflies attain *Leishmania donovani* parasites when they feed on an infected mammalian host in search of a blood meal. The protozoan has two forms, termed amastigote – round, nonmotile, and only 3–7 micrometers in diameter. The amastigote forms of the parasite taken up by sandflies are not regularly developed in the peripheral circulation; rather they exist in the skin itself. Amastigotes are intracellular parasite found in the phagolysosomes of macrophages and other phagocytes, and their uptake by the bloodsucking sandfly is aided by the acerbic action of the mouthparts (Lane 1993). Taken into the stomach of the sandfly, the amastigote swiftly transmutes into a second *L. donovani* form, identified as promastigote (Chakrabarti et al. 2013). This form is spindle shaped, triples the size of the amastigote, and has a single flagellum that allows for motility. The amastigotes in the gut of sandflies differentiate into promastigotes and grow. Then the *L. donovani* multiplies and further differentiates into other stages, metacyclic promastigote being the final mammalian-infective stage which moves to the foregut of the sandfly (Rogers et al. 2002a, b). Promastigotes subsist extracellularly in the sandfly's alimentary canal, replicating asexually, and then migrate to the proximal end of the gut where they become controlled for a spewing transmission.

The epidemiology of VL rests on the coexistence and interface of the parasite, the vector, and the host population, exaggerated by the variations in both biotic and abiotic circumstances (de Almeida et al. 2011). A host plays as a reservoir host if it can transmit the parasite into the next stage that into the vector. The local microgeographical physiognomies of the transmission sites, the present and past acquaintance of the anthropological characteristics to the parasite, and the extensively changing human behavior regulate the infection status in human. Human driving forces alter the local ecology that influences the inflow of parasite transmission and possible reservoir hosts from forest-rural areas into semiurban and urban human settlement. Humans may bury the sylvatic cycle in search of agricultural land, human settlement, timber manufacture, road creation, supplementary economic assistances in the jungle, and other enzootic areas (Feliciangeli et al. 2006). The disease transmission inclines to be high in agrarian hamlets where houses are normally built with earthen walls and mud floors and cattle and other livestock are kept adjacent to human residences (Bern et al. 2010). Keeping domestic animals inside the houses would increase human-sandfly contact, hence favoring the transmission of disease. Subsequently, migration of people from rural to urban environments is a prime reason for the establishments of VL in periurban and urban settlements (Aagaard-Hansen et al. 2010).

1.5 Symptoms of Visceral Leishmaniasis (Kala-azar)

Kala-azar (KA) is a fatal form of leishmaniasis with a mortality rate of 75–95%. Macrophages affected by the parasite spread the infection throughout the body, and patients develop pancytopenia and immunosuppression (Oryan and Akbari 2016). Symptoms of KA infection can take many different and diverse forms. The incubation period of KA varies between 10 days and 10 years, the average being 6 months (Manson-Bahr and Apted 1982). A multitude of clinical symptoms ensue progressively, the most important being splenomegaly, recurring and irregular fever, anemia, pancytopenia, weight loss, and weakness (Luo and Levitt 2008). Hemoptosis and other bleeding disorders, diarrhea, and cough may be present. Affected patients become gradually more anemic, weak, cachectic, and susceptible to intercurrent infections. The enlargement of the spleen and liver are the main signs of KA. The disease is a silent killer, perpetually carnage almost all untreated patients (Boelaert et al. 2009). Hyperpigmentation, perhaps directed to the name KA (black fever in Hindi), has only been shown in visceral leishmaniasis (VL) patients from the Indian subcontinent, but nowadays this sign is rare and was possibly a feature of lengthy sickness in the era when effective treatment was not accessible (Chappuis et al. 2007). KA affects not only the feeblest in the society, for example, children and those weakened by other sicknesses like HIV and tuberculosis, but also healthy adults and the poorest social groups. Incubation period of VL varies between 2 weeks and 2 years.

1.6 Spatiotemporal Distribution of Visceral Leishmaniasis

Information on global occurrence of VL is disparate and sparse. Leishmaniasis occurs in 88 countries in tropical and temperate regions, of which 72 are least developed countries, viz., Africa, Asia, the Mediterranean Basin, Southern Europe, and Central and South America. However, the distribution of VL is nonuniform; it is inconsistent and often linked with areas of drought, famine, and thickly populated villages with little or no sanitation. Approximately 147 million people are at risk of VL in the southeast region (WHO 2016), but the true picture remains largely veiled since a substantial number of cases were never recorded. The disability-adjusted life years (DALY) burden was 2,357,000, and the total deaths were 59,000 per year (WHO 2002). The annual estimate of the incidence and prevalence of KA cases worldwide is 0.5 million and 2.5 million, respectively, and of these, 90% of the cases occur in India, Nepal, Bangladesh, and Sudan (Bora 1999). Postkala-azar dermal leishmaniasis (PKDL) is prevalent in India, Sudan, and Kenya. Depending upon the eco-epidemiological conditions, leishmaniasis can present sylvatic transmission cycles. Among the most important factors composing those conditions, bio and abiotic factors as well as parasite, vector, and host species are involved.

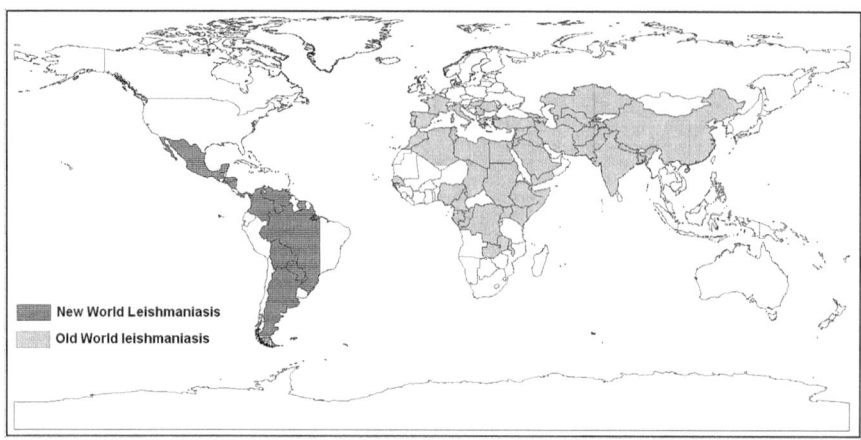

Fig. 1.5 Distribution of leishmaniasis in the world. (*Source*: Bhunia 2014)

1.6.1   Global Status of Visceral Leishmaniasis or Kala-azar

Geographically, the disease is distributed globally between tropical and subtropical regions (Aversi-Ferreira et al. 2015). As per the 1957 survey reports, leishmaniasis was existing in all the areas except for Oceania. It was observed that 6 of the 46 countries in the Caribbean and Central and South America, 10 of 58 countries in Africa, 8 of 18 countries in Southwest Asia, and 7 of 24 countries in South-Central and Southeast Asia documented this disease (Fig. 1.5). In Central America, especially in Costa Rica, Honduras, and Nicaragua, previously isolated VL cases were recorded. In Brazil, VL is distributed widely in the south, east, and central regions of the country. The disease is highly endemic in the states of Bahia and Ceara, which together account for 70% of the total cases of VL in Brazil (WHO expert committee report 1991). Up to 1989, 15,000 cases of VL had been recorded in the states of Alagoas, Espirito Santo, Goias, Mato Grosso do Sul, Minas Gerais, Para, Paraiba, Pernambuco, Piavi, Rio Grande Norte, and Sergipe including Bahia and Ceara (WHO expert committee report 1991). According to WHO, leishmaniasis is one of the seven most vital tropical diseases, and it epitomizes a serious world health problem with an extensive spectrum of clinical appearances with possibly fatal outcomes (Andrade-Narvaez et al. 2003).

There are approximately 98 countries and territories with 1.5–2 million new leishmaniasis cases documented each year. In the 1990s, HIV/VL co-infection and universal global warming enhancing the probable habitat of the sandfly led to replication the amount of cases from 1987 to 2014 despite developing medical technology. Handler et al. (2015) reported that each year around 400,00 people are having VL with a mortality rate of 10%, going up to 20% in some regions. In 2017, 20,792 out of 22,145 (94%) new cases reported to WHO occurred in seven countries: Brazil, Ethiopia, India, Kenya, Somalia, South Sudan, and Sudan (Fig. 1.5). The disease burden connected with VL, measured in DALYs, was estimated to be

1,980,000 (1,067,000 for male and 744,000 for female) in the year 2000. VL is caused by *L. donovani* in the Indian subcontinent, Asia, and Africa (in adults and children) (Murray et al. 2005), by *L. infantum* in the Mediterranean Basin and Southwest and Central Asia (Croft et al. 2006), and by *L. chagasi in* South America (Salomon et al. 2006). In Sudan, for example, a major decade-long epidemic of VL occurred from 1984 to 1994 (Seaman et al. 1996). Some studies estimate that the disease caused 100,000 deaths in a population of around 300,000 in the western upper Nile area of the country (Seaman et al. 1996).

In Latin America, the number of VL cases has increased in northern Argentina, Brazil, Paraguay, Venezuela, Uruguay, and North America. The number of reported cases in Brazil is comparatively higher and distributed in both rural and urban areas (Aversi-Ferreira et al. 2015). Since the last decades, the disease is sparsely distributed in subSaharan Africa. However, the number of cases is increased in the eastern African region. Countries like Sudan, Ethiopia, Somalia, Uganda, Kenya, and Eritrea have the most number of infections in relation to VL transmission (Aversi-Ferreira et al. 2015), and very few cases were reported from France, Portugal, Russia, and China (Alvar et al. 2012). In the Mediterranean Basin (Albania, Morocco, Greece, Italy, Spain, Tunisia, and Algeria) have reported of VL cases. In the Asian subcontinent (Middle East and Central Asia), a significant number of VL cases were reported in Iran, China, and Georgia. In South-Central and Southeast Asia, kala-azar is endemic in India, Bangladesh, and East Pakistan wherein the past extensive epidemics have occurred. Apparently, the disease is absent for the most part from Burma, Vietnam, Laos, Thailand, Malaya, and Indonesia. In Bangladesh, VL was significantly eliminated between 1953 and 1970, possibly due to mass chemotherapy with pentavalent antimonials and extensive insecticide spraying to mitigate malaria. Following the end of the malaria control program in 1970, sandfly densities increased and so did the cases of VL which presently emerge at a rate in excess of 15,000 per year (Al-Masum et al. 1995). In Pakistan, 239 cases of VL due to *L. infantum* were recorded between 1985 and 1995; of these, 52% were children less than 2 years and 86% were children less than 5 years, symbolizing a rise of tenfold in infantile (Rab and Evan 1995). VL has been recognized to subsist in the Himalayas in Pakistan for over three decades. Nevertheless, newly sporadic cases may come out in the North-West Frontier Province (NWFP), Punjab, and Azad Jammu and Kashmir (AJK). All of these areas are mountainous and enclosed by great farming communities (Rab and Evan 1995). In adjoining India, the main endemic area of VL is Bihar, West Bengal, Assam, Orissa, and the eastern part of Uttar Pradesh (Fig. 1.6). One of the biggest epidemics took place in 1978 in North Bihar where over half a million people fell victim to VL. In the first 8 months of 1982, 7500 cases were documented in India, and between 1987 and 1988, 22,000 cases of VL were recorded (WHO expert committee report 1991). Presently, KA has been perceived with the growing occurrence in patients who have AIDS or who are venous drug users or both, signifying a possible transmission mechanism through contaminated needles (Torres-Guerrero et al. 2017).

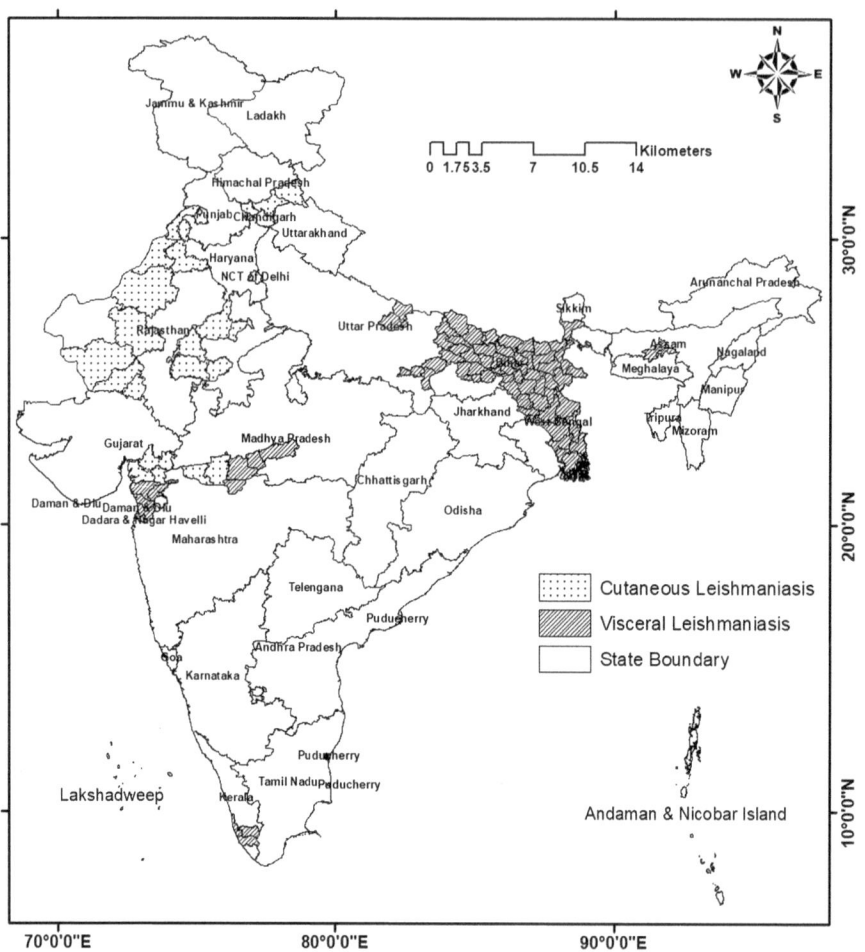

Fig. 1.6 Geographical distribution of leishmaniasis in India

1.6.2   Regional Status of Visceral Leishmaniasis or Kala-azar

The Indian subcontinent (ISC) has historically suffered most of the global burden of VL. The disease is pure anthroponosis in India. Humans appear to satisfy all the condition of a reservoir of natural infection, i.e., close association with the vector *Phlebotomus argentipes*. In the ISC, the endemic belt used to extend from Meghalaya, Assam (Brahmaputra Valley), Bihar, West Bengal, Uttar Pradesh, coastal Odisha, coastal Andhra Pradesh, and Tamil Nadu. The epidemic is largely concentrated in four Indian states – in the northeastern tip, predominantly in Bihar state, where occasional epidemics occur (Sanyal 1985). Recently, few cases of VL were reported from coastal Kerala (Fig. 1.6). KA was prevalent in Calcutta prior to 1835 (Twining 1832) and has spread from North Bengal to Assam (Rogers 1897).

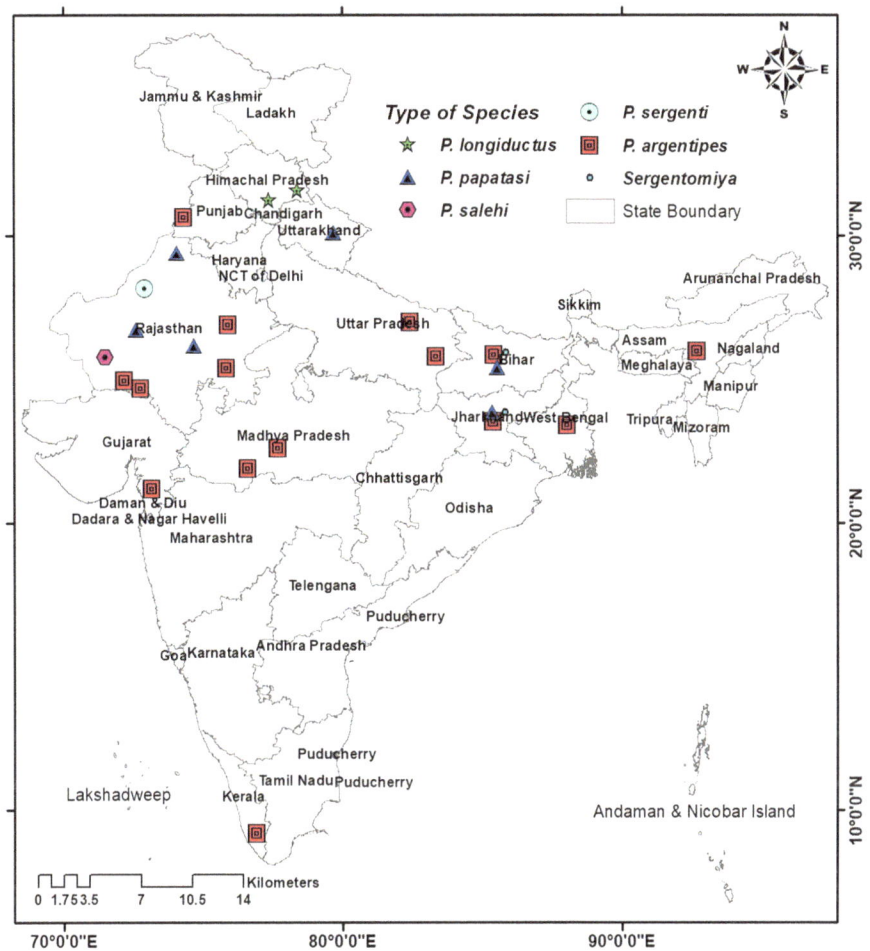

Fig. 1.7 Geographical distribution of vector species of leishmaniasis in India

The first outbreak was recorded from Garo Hills in the late 1860s and 1870s (Clarke et al. 1996). Later, the outbreak of the disease commenced in Bihar, Assam, West Bengal, and Tamil Nadu in 1930, 1940, 1943, and 1947, respectively. The disease is also found in Jharkhand, Uttar Pradesh in decreasing order of endemicity as one proceeds westward and in Odisha (Fig. 1.6).

However, since 2011 there has been a substantial drop in the number of cases in the ISC. As per the WHO elimination goal, the diminution in the prevalence of symptomatic VL to achieve under 1 case/10,000 people/year at subdistrict level in the ISC by 2020 (Jervis et al. 2017). Around 600 Phlebotominae species are recognized, most of them belonging to the genera *Phlebotomus* in the Old World and *Lutzomyia* in the New World. Only 30 of the species are vital in respect to public health, and 10% of the species act as a disease vector. *Phlebotomus* (Palearctic genus) is distributed in Europe, Asia, northern Africa, central parts of the Arabian Peninsula, and other regions of the Old World. *Lutzomyia* is the main genus of North America (Belen and Alten 2005). Generally, *Phlebotomus* is found in arid and semiarid area; and some species prefer peridomestic and indoor situations (e.g., human dwelling, organic waste, dung, feces, rodent holes, leaf litter, cracks, and crevices in the walls). The spatial distribution of *Phlebotomus argentipes* in India is illustrated in Fig. 1.7. *Lutzomyia* transmits the disease in forest habitation. However, the distribution pattern of sandflies and leishmaniasis seems to be altering (Belen and Alten 2005).

1.7   The Problems

A combination of factors has seen the resurgence of VL as a major disease throughout the world. The recent experience in tropical and subtropical countries emphasizes the scope of the problem. The geographical distribution of sandflies (*P. argentipes*) and their life history, hosts, and ability to transmit the infection are determined by both intrinsic and extrinsic factors. The intrinsic factors are the biochemical and physiological properties of the sandflies that determine its reaction to the external conditions of its habitat. The extrinsic factors are the abiotic and biotic components of the habitat that influence the sandfly biology. VL is zooanthroponosis in nature. Moreover, the vector repetitively infects all persons present in its vicinity. The distribution of the vector is highly influenced by meteorological and eco-environmental factors of an area. The range of values of annual mean minimum temperature (16–20 °C) and maximum (25–27.5 °C) temperature, relative humidity (80–90%), precipitation (1000–1550 mm), and elevation (<300 m from MSL) are significant for the spread of *P. argentipes* breeding places. The presence of inland surface water bodies favors the breeding and propagation of immature stages of sandflies. Extremes of climate affect the sandfly population adversely. Presence of crop land in the area is generally characterized by clay and alluvial type of soil. The land provides moisture to the soil in the adjoining domestic biotope. Sandflies require a sugar meal, taken from plant material; therefore, they would expect to be less abundant in areas with little or soft stem vegetation. Sugarcane, banana, and tobacco are the main cash crops in these regions. Majority of the sandflies may be found in mud-plastered and thatched houses and recovered from the human dwellings and cattle sheds. *P. argentipes* is essentially a "wet season" species, and with the onset of rains, there is generally a sharp upswing in the density of species. Though sandflies are incapable of flying long distance and move about by their characteristic hopping movement. Vertical distribution of *P. argentipes* has been reported to be up to 6 feet on the wall, and they have been encountered up to a distance of 270–505 meters from their breeding places. Peak density was observed soon before and after monsoons during March–April and August–October, respectively. Therefore, deducing social, economic, and environmental impacts of the disease requires in-depth knowledge of pathogen, host, and vector biology and ecology.

The results of an extensive microhabitat survey of endemic sites and the relationship between vector vis-à-vis KA in relation to environmental and geographical factors are based on these factors.

1.8   About This Book

Infection transmitted by sandflies causes significant rates of death and disease, especially in developing countries. *Spatial mapping and modelling for kala-azar disease* reveals how using geographic information systems (GISs) can provide a greater understanding of KA are spread and explores the use spatial mapping and modelling in KA monitoring, management and control. The text provides readers with better understanding of open GIS consortium of public health problem analysis and modelling. Utilizing simple illustration based on real data, it will offer step-by-step instruction on developing KA risk mapping and modelling at different spatial and temporal scales. The book also provides the role of telemedicine, Internet of Things (IoT), data science, and cloud computing technique to formulate disease control program. This book:

- Provides an overview of KA, GIS-based mapping and modeling, and availability and importance of accurate epidemiologically relevant geographical data
- Demonstrates and simulates the prevalence of KA around the world
- Analyzes microgeographical factors associated with the KA distribution
- Summarizes some key spatial techniques and how they can be used to aid in the analysis of public health data
- Illustrates open GIS consortium for public health data analysis and mapping
- Provides geospatial technique-based case studies for modelling the future potential distribution of kala-azar risk mapping and vector breeding site demarcation

1.9   Conclusion

Visceral leishmaniasis (VL) or kala-azar (KA) is still a big public health challenge. Leishmaniasis is reported in 89 countries in tropical and temperate regions, of which 72 are least developed countries. Ninety percent of the VL are found in Bangladesh, India, Nepal, Sudan, and Brazil. Globally reported confirmed VL cases have decreased since 2011. The earliest epidemic of KA was first documented in 1824 and was named Jessore fever as it was observed in the Jessore district of India. *Leishmania donovani* is responsible for anthroponotic transmission, and *Leishmania infantum* is accountable for zoonotic transmission. Around 600 *Phlebotomine* species are identified globally – *Phlebotomus* in the Old World and *Lutzomyia* in the New World. Mackie (1914) first suspected the possible association of the genus *Phlebotomus* with the transmission of KA. The disease is transmitted by the bite of infected female sandfly, *Phlebotomus argentipes*, in India and Southeast Asia. Despite the expansion in scientific acquaintance and medical sciences, there is still an essential for fast and cheap recognition of *Leishmania* infections especially in endemic areas. Hence, in the future, the researches appraising the antileishmanial activity of numerous products or chemically improved compounds are required to find new prospects in fruitful treatment of VL infections.

Chapter 2
Geoinformatics and Kala-azar Disease Transmission

Abstract Geoinformatics is concerned about georeferenced data input, storage, recovery and addition, image processing operations, spatial analysis tools, visualization, plotting, and graph in a systematized form. The spatial distribution of the disease and vectors encompasses the practice of computational investigation and illustration of geographic data using the so-called geoinformatics. Moreover, several environmental variables derived from satellite data such as climate, land use/land cover, and other environmental aspects that influence the activity of pathogens, vectors, and their interactions with hosts and reservoirs can be used for mapping and monitoring the disease distribution pattern. Subsequently, the geographically referenced data may aid in numerous aspects, like documentation and spread of disease over time, population clusters at risk, forms of disease epidemics, ability accessible to healthcare and program intercession planning, and determination in disease outbreak. The Global Navigation Satellite System (GNSS) allows the correlation of the geographical distribution of VL with environmental factors. Hence, geoinformatics is a powerful tool for disease surveillance, envisaging its epidemics and monitoring control program.

Keywords Remote sensing · GIS · GNSS · Disease mapping · Environmental variables

2.1 Introduction

Kala-azar is a spatially determined phenomenon, and its identification of spatially risk factors plays a crucial role in prophecy, prevention, and control. Since Hippocrates circa 400 B.C., "an endemic disease is determined by the nature of a certain place." Hippocrates' ecological notion of the tropical disease was introduced by Galen in the early Christian era and conceded untouched through innovation of science during the Renaissance. In 1966, Pavlovsky established the theory of

G. S. Bhunia, P. K. Shit, *Spatial Mapping and Modelling for Kala-azar Disease*,
SpringerBriefs in Medical Earth Sciences, https://doi.org/10.1007/978-3-030-41227-2_2

natural nidality of vector-borne disease which is based on three basic aspects – (i) diseases tend to be restricted spatially; (ii) the spatial variation depends upon the physical and biological conditions that support pathogen, vectors, and reservoirs; (iii) concurrent and forthcoming variation in disease pattern can be determined via biotic and abiotic aspects. Visualization of geographical distribution of data consents detaining present patterns and investigation of disease process. The geographical distribution of the disease and vectors comprises the usage of computational investigation and illustration of geographic data using the so-called geoinformatics or spatial analysis tools. Geoinformatics is composed of georeferenced data input, data storage, retrieval and integration, image processing functions, spatial analysis tools, visualization, plotting, and graph in an organized form.

Geoinformatics have distinct benefits over conventional ground measurements as they can assemble information recurrently and inevitably. Before 1970, Cline emphasized the scenarios of using aerial picture and other RS techniques in epidemiological research (Cline 1970). In 1971, the National Aeronautics and Space Administration (NASA) established the Health Applications Office (HAO) to unify and coordinate collaborations and programs on smearing RS techniques in disease research, and considerable revolutionary work was commenced in the late 1970s. On July 23, 1972, the first Earth Resources Technology Satellite (ERTS-1) was launched effectively, beckoning a historical innovation in technology to put on space technology in human health research (Zhang et al. 2013). In 1998, the National Aeronautics and Space Administration's (NASA) Center for Health Applications of Aerospace Related Technologies (CHAART) appraised current and planned satellite sensor systems as a first step in allowing human health scientists to regulate data pertinent for the epidemiologic, entomologic, and ecologic features of their research, in addition to developing RS-based models of transmission risk (Beck et al. 2000). RS brings new paraphernalia for disease monitoring, exploring their connotation with numerous sensed parameters and eventually generating an intelligent alert system. Satellite data provide information of the earth objects or phenomenon based on reflected, emitted, and backscattered radiation. However, there has been increasing usage of satellite data in large-scale environmental mapping and modelling in relation to disease occurrence. Image processing operation was carried out to compute the spectral indices that are sensitive to environmental factors such as vegetation vigor, water bodies, surface temperature, etc. This information will be helpful for identifying the suitable location of specific vector and host habitats associated with the disease transmission. Thus, the use of satellite-derived information can help to improve disease risk maps by providing novel sources of environmental information that are relevant to vector and host ecology. Several literature of geospatial technology applications in health have previously been commenced and focused on specific types of infectious disease. The epidemiological and entomological research on KA that used Earth Observation data in mapping, modelling, and forecasting are critically describe focused research question. The Scopus, PubMed, J-GATE@ERMED, JCCC@ICMR, and Web of Science databases were searched electronically to attain relevant literature and articles. Boolean operators coalescing several keywords relevant to the KA research were inquired in the abovementioned databases. The keywords were "Kala-azarLit," "Visceral Leishmaniasis," "Remote Sensing and GIS," "Satellite data," "Environment," "Climate," "HealthGIS," "Risk zone Model," "Prediction," and "Spatial analysis." The database search was limited to studies reported in English language journals and indexed with the keyword strategy. Literatures addressing *kala-azar*, *visceral leishmaniasis*, *leishmaniasis*, *RS and GIS* methodology, or human health-related geospatial appliances were filtered, and the duplicates were removed. In addition, it draws extensively on literature culled manually and compiled over the years in relation to the authors' specific work on *leishmaniasis*.

2.2   Remote Sensing (RS) and Visceral Leishmaniasis (Kala-azar) Transmission

Remote sensing (RS), in a universal logic, is the little or large-scale attainment of data on an object or phenomenon using recording or real-time sensing devices that are not in physical or direct contact with the object or phenomenon being studied (Campbell-Lendrum et al. 2002). While RS is now decisively related with satellites, all the ideas and the technology were mainly distinguished a long time before satellites came into use for this persistence. RS usually denotes the use of aerial/spaceborne sensor technologies to perceive and categorize objects (both on the earth surface and in the atmosphere and ocean) based on the electromagnetic radiation emitted from it. Based on the spectral characteristics, a lot of information can be obtained from the earth surface features (e.g., vegetation, land use/land cover (LULC), water bodies, etc.) which are closely related to environment-related disease and can influence their dispersal accordingly. Satellite data were started to elicit the landscape and climatic information that could be suggestively providing information to disease modelling, referring to predictive mapping, spatial distribution and propagation of the pathogen and vectors, health risk analysis, understanding the transmission dynamic, demarcation, execution of suitable control strategies, and their evaluation.

Miranda et al. (1996) have used SR-satellite imagery technology on an epidemiological survey with American cutaneous leishmaniasis. Lima et al. (2002) investigated the geographical distribution of human TL cases and correlated with the occurrence of the remaining vegetation and water streams through satellite monitoring. Bern et al. (2005) investigated the spatial pattern and risk issues for anthroponotic VL through active case detection and serologic screening in Bangladesh. Bhunia et al. (2010a, b) studied about various environmental parameters (e.g., temperature, humidity, rainfall, altitude, vegetation condition, and land use characteristics) derived through remote sensing data for correlation with the distribution of KA. Bhunia et al. (2010a) conducted a correlative study with altitude, temperature, rainfall, humidity, and normalized difference vegetation index (NDVI) extracted from SRTM and MODIS satellite data with the distribution of KA in the northeast corner of the Indian subcontinent. Kesari et al. (2011) analyzed the vegetation vigor and land use characteristics of endemic and nonendemic areas of KA in India using IRS-1D Linear Imaging Self-Scanning (LISS)-III sensor data. Bhunia et al. (2011a) carried out a correlative study between VL distribution and surface water bodies extracted from the Landsat Thematic Mapper data. Bhunia et al. (2012a) investigated the role of LULC derived through Advanced Very-High-Resolution Radiometer (AVHRR), Moderate Resolution Imaging Spectroradiometer (MODIS), Thematic Mapper (TM), and Linear Imaging Self-Scanning (LISS)-IV for leishmaniasis transmission resulting in a framework highlighting the links between LULC and areas endemic for KA. Bhunia et al. (2012a, b, c) reported that seasonal variation of maximum and mean NDVI derived through Landsat TM satellite data were strongly associated with KA. Bhunia et al. (2012a, b, c) suggested a geoenvironmental approach to discriminate the high- and low-risk areas of KA in India. Golpayegani et al. (2018) studied the relationship between environmental factors (vegetation and elevation) and the prevalence of disease in Iran.

Although large numbers of earth observation (EO) data are available till date, very few have been found in KA research. By far the most important satellite data such as Landsat's Multispectral Scanner (MSS) and Thematic Mapper (TM), Indian Remote Sensing (IRS)-s Linear Imaging Self-Scanning III and IV, the National Oceanic and Atmospheric Administration (NOAA)'s Advanced Very-High-Resolution Radiometer (AVHRR), and France's Système Pour l'Observation de la Terre's (SPOT) CBERS 2/CCD have been used. The situation is more challenging when fine spatial, spectral, and temporal resolutions are essential, and the only solution is to use two or more causes of remotely sensed imagery. However, each satellite data set has a tendency to a number of individual flukes, which can make direct comparison difficult. The most important parts of the EM spectrum are visible and the near infrared, as they examining reflection properties in the bands it is possible to discriminate land cover classes. The discrepancy of spectral response with the angle of the sun creates image interpretation as much art as a hard science. Most of the studies have used satellite data, while some additionally incorporated in situ data. For example, vegetation indices are important factor that showed strong correlations with the KA vectors' behavior and their biological cycle. Hence most of the research works were considered the satellite-derived vegetation index in combination to the data to model spatial and temporal dynamics of *Phlebotomus* species and VL. Subsequently, topography is also a significant factor in the transmission of VL, as it affects the local weather condition, undulation pattern of the landscape, and living conditions of the people (Bhunia et al. 2010a, b).

2.3 Remote Sensing of the Vector Environment

Based on the eco-epidemiological conditions, the VL can present sylvatic or domestic transmission cycle (Carreira et al. 2015). The eco-epidemiological features are significantly prejudiced by the wide dispersal of the parasites, the presence of a great diversity of vector species, as well as the existence of local environmental aspects influencing the population of human hosts, vectors, and reservoirs (Maia-Elkhoury et al. 2008). Presently, there has been a growing interest in assimilating RS-derived variables within the progression spatial data modelling to recognize high-risk areas of KA (Bhunia et al. 2014; Bhunia and Shit 2019). The vector abundance is used to evaluate disease risk, but there are discrete inconsistencies between risk and disease incidence. The role of RS data in vector propagation involves retrieving environmental variables that proliferate the vector ecosystem such as land cover, temperature, humidity or vapor pressure, and precipitation. Elnaiem et al. (1998) investigated the association between environmental variables (rainfall, air temperature, land surface temperature, soil, vegetation) derived through remotely sensed data and *P. orientalis* distribution, and found a significant association with the *Acacia seyal* and *Balanites aegyptiaca* tree species, black cotton soil of eastern Sudan. Rogers et al. (2002a) investigated the link between satellite-derived seasonal environmental variables and vector biology. Peterson et al. (2004) used genetic algorithm for rule-set prediction (GARP) for ecological niche modelling to understand the geographic distribution of vector species. Sudhakar et al. (2006) reported that highly succulent crop (sugarcane, bananas, etc.) and local prevailing conditions influenced *P. argentipes* density in Bihar (India). Rossi et al. (2007) reported the distribution of *P. perniciosus* in relation to environmental features in southern Italy and stated that vector abundance is significantly higher in green vegetated environments in the coastal side. Ready (2010) investigated the role of climate change on leishmaniasis distribution as the environmental variation influences the seasonal abundance of sandfly species. Bhunia et al. (2011a, b) reported that *P. argentipes* were found to be associated with the distance from open water and particularly abundant near nonperennial river banks, whereas its association with the perennial rivers was focused further away from water source. Bhunia et al. (2012a, b, c) analyzed the correlation between *P. argentipes* density and minimum, maximum, and mean NDVI value and reported that mean and maximum NDVI values were strongly correlated with the seasonal abundance of *P. argentipes*. Most of the RS-related studies associated with VL have been space-based static researches and have not been considering the attribute of time, which may be vital for visualizing and predicting works. Distribution of land use/land cover (LULC) characteristics was studied by a probabilistic approach to identify a set of "suitability estimates" depending on the probability of *P. argentipes* presence (Bhunia et al. 2012a, b, c). Jeyaram et al. (2012) presented the development of an early warning system based on linear and nonlinear model for predicting *P. argentipes* density in India. Kesari et al. (2013) suggested that remotely sensed environmental variables (minimum LST, mean LST, and mean RDVI) are the best environmental covariates for predicting *P. argentipes* density in Bihar (India).

Most of the research work considered multispectral remote sensing data in sandfly distribution and breeding habitat determination. Very few studies used microwave, thermal, and hyperspectral satellite imagery to analyze the environmental variables for monitoring sandfly in a straight line from space, but they can be used to recognize favorable environment or breeding places. Remote sensing

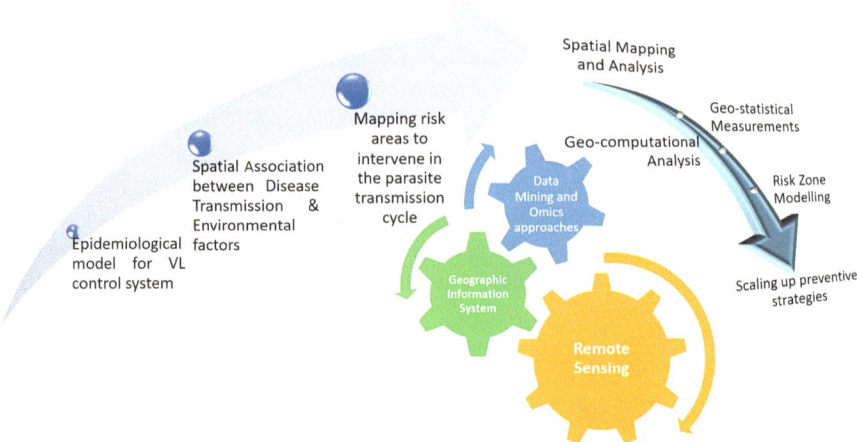

Fig. 2.1 Categorization of kala-azar research using remote sensing and GIS application

technologies, which allow mapping of environmental variables, have already been used in different epidemiological studies (Bhunia et al. 2010a, b) but so far rarely deal with visceral leishmaniasis. Few studies are available that include the extraction of environmental indicators like meteorology, vegetation, altitude, etc. (Sudhakar et al. 2006). Nieto et al. (2006) developed an ecological niche model to delineate the distribution and potential risk zone of visceral leishmaniasis. Figure 2.1 shows the remote sensing and GIS application in KA research.

2.4   Geographic Information System (GIS) and Kala-azar

According to Bill (1999), a "Geographic Information System (GIS) is a computer-supported system comprising of hardware, great complexity of software, data and the subsequent spatial analysis." GIS seems to be an information system that turns geographical data into evocative mapped output. In public health and epidemiology, GIS has a long history. Maps demonstrating the spatial distribution of yellow fever were made in 1798, and in the 1850s, John Snow (an English anesthesiologist) plotted the places of cholera cases in Soho, London, leading him to the inference that water from the extensive street pump was accountable for the cholera epidemic (Snow 1849). Since 1993, WHO's public health mapping program has been initiated to endorse and implement GIS into decision making for a wide variation of infectious disease. GIS and methods of spatial analysis have been extensively used in spatial distribution of VL since 1990. The Public Health Mapping Program-based WHO Communicable Diseases has been established with the aim of offering better access to meek low-cost spatial information and interrelated data management and mapping systems to public health planner at all levels of the health system (Fradelos

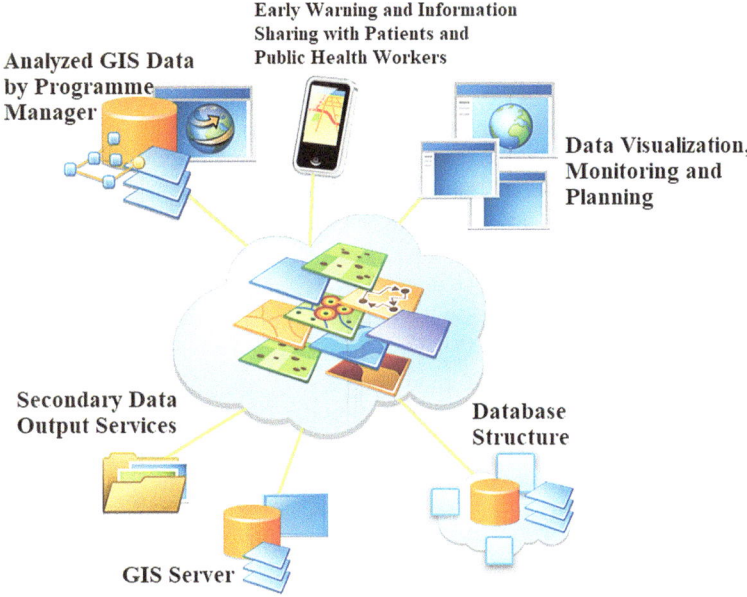

Fig. 2.2 Application of GIS in kala-azar control program

et al. 2014). With the use and exploration of data obtained by contemporary GIS technology, it is probable to attain pertinent information on the site of vector to calculate the vector density and pay ideal techniques to generate a rational environment-friendly approach to control the nuisance sandfly and kala-azar transmission (Fig. 2.2). Additionally, GIS techniques and regression models have been used in several studies to generate disease risk maps (Bhunia et al. 2012a, b, c; Kesari et al. 2013). Mandal et al. (2018) used the spatial statistical tool to determine the trend in spatiotemporal dynamics in VL-affected villages in India. Mandal et al. (2019) appraised the residual abilities and the intervention effects of dichloro-diphenyl-trichloro-ethane (DDT)- and synthetic pyrethroids-indoor residual spray (SP-IRS) data at the household level under the national vector control program; the spatial risk map and sandfly density analytical model based on the housing characteristics have been prepared through GIS. There are several uses of GIS in kala-azar studies: (i) defining spatial distribution of disease, (ii) investigating spatial and temporal trends, (iii) mapping population at menace, (iv) stratifying risk aspects, (v) evaluating resource allocation, (vi) forecasting and targeting intercessions, and (vi) monitoring disease and interventions over time. GIS applications related to kala-azar have been introduced and used in, for example, quantifying environmental hazards and their influence on public health (Salomon et al. 2006), for policy and planning purposes (Clements et al. 2006), for the surveillance and monitoring of diseases (Kalluri et al. 2007), in environmental health (Bhunia et al. 2010a, b), and for spatial mapping and modelling (Bhunia et al. 2011a, b), risk zone mapping (Bhunia et al. 2012a, b, c), the development of an early warning system (Jeyaram et al. 2012), and spatial clustering (Bhunia et al. 2013).

2.5   Global Navigational Satellite System (GNSS) and Visceral Leishmaniasis

The Global Navigation Satellite System (GNSS) provides independent geospatial positioning with global coverage. GNSS's satellites are ensembles that permit any user at or near the Earth to govern their location with a precision from meters to centimeters. Up to 2013, NAVSTAR (US) and the Global Navigation Satellite System (GLONASS) of the Russian Federation are the only operational GNSSs. After that, Galileo (European Union's positioning system), Indian Radio Navigation Satellite System (IRNSS), QZSS (Japan), and COMPASS (China) are used to detect environmental changes, which have a significant effect on local population health, satellite communication for medical endeavors, and advancing medical knowledge through space medicine programs. GNSS system consists of a constellation of satellites that send a continuous signal to the Earth. Individuals wanting to use GNSS to regulate their location must have an antenna that obtains the signals imminent from the satellites and a receiver that decodes these signals. GIS together with GNSS is used in many vector control activities in various countries, by identifying spatial location of infected persons. GNSSs allow the correlation of the geographical distribution of VL with environmental factors. Earlier several research works on VL have been used Global Positioning System (GPS) to collect the coordinates of affected household of VL to visualize the spatial pattern of disease (Chapman et al. 2018), coordinates of the individual household of location of VL (Abdullah et al. 2017), and validation of risk map (Tsegaw et al. 2013). Additionally, distance to the health facility will also be used to conduct random sampling frames for field data collection, mapping households, tracking, model validation, correlative study, and emergency service of kala-azar (Fig. 2.3).

Fig. 2.3 Application of GNSS in VL control strategy

2.6 Conclusion

The gradual climate change due to global warming modifies ecology, habitats, hosts, agents, and vectors in endemic areas. Further, this disease may swell the physical range, and new epidemic and pandemic of these diseases may happen in areas formerly free of the disease. The beginning of RS technology with its synoptic coverage, high repetivity and cost effectiveness may support for the habitat's surveillance, vector densities and prediction of disease risk areas, has opened up new vistas in the epidemiology of kala-azar. Innovative research on spatial analysis is required to be built as structure of theories, models, technologies, and novel methods of depiction that go beyond previous GIS models, giving birth to innovative practices for addressing ambiguity. Even with the important research efforts using static and dynamic models for the VL epidemic, research breaches still exist. Research focus should consider the following: (i) sophisticated geospatial model to determine underlying VL transmission dynamics; (ii) integrating real-world data for model validation and verification; (iii) simulation test of the predictive tool to confirm the feasibility of the model and control; (iv) creation of latent risk area maps rely on the circumstances promising to vector propagation of exotic diseases; (v) expansion of early warning system before endemic; (vi) enhanced deterrence initiative directing ingress of risk; and (vii) enhancement in bioterrorism event data management. However, the remote sensing and GIS-integrated spatial map and model related to VL transmission remain an active and study-worthy area that will help public health organizations to understand and prevent the disease. Subsequently, high-level applications to be initiated by open source data and the ability of using analytical tools and display capabilities from different sites resulting in complex data manipulation.

Chapter 3
Microgeographical Factors of Kala-azar Disease

Abstract Changing landscape and climate change can significantly affect local climate and environmental aspect more accurately. The spatial distributions of microgeography and habitat loss are the current challenges to humankind also influencing the vector ecology and vector breeding habitats. Microgeographical factors like temperature, relative humidity, rainfall, soil, peridomestic vegetation, topography, natural vegetation, deforestation, surface waterbody, land use/land cover, population density, housing characteristics, family size, illiteracy rate, unemployment, urbanization, population migration etc. played an important role in kala-azar transmission. This chapter demonstrates the role of remote sensing data to extract the geographical factors. This chapter also illustrates two examples by identifying some microgeographical factors that have shaped kala-azar propagation in the past and those that appear to be playing a part today.

Keywords Microgeography · Satellite data · Environmental variables · Demographic variables

3.1 Introduction

To understand the cause of epidemic and the transmission of disease, knowledge on geographical and environmental aspects are taken into account. Visceral leishmaniasis (VL) or kala-azar (KA) is primarily associated with the tropical and equatorial zones, where the parasite and vectors are close to the reservoir and hosts, favoring *Phlebotomine* reproduction (Fig. 3.1). In terms of geographical factor, VL is limited to the geographic area where the vector inhabits and climate is also promising for

© The Author(s), under exclusive license to Springer Nature Switzerland AG 2020   29
G. S. Bhunia, P. K. Shit, *Spatial Mapping and Modelling for Kala-azar Disease*,
SpringerBriefs in Medical Earth Sciences, https://doi.org/10.1007/978-3-030-41227-2_3

transmission of disease (Aversi-Ferreira et al. 2015). However, there are some places and certain environmental factors that are comparatively encouraging to the transmission of disease, also associated with the high abundance of vector and reservoir; these are considered as local issues that would be responsible for the frequency of recurrence of the disease. *Phlebotomine* sandflies principally reside in hot and wet tropical climates with regular pluvial index (Rebelo et al. 2001). However, dry and hot places are also reported for the presence of sandflies in some published literature. And the distribution of sandflies is dependent on the environmental variables such as temperature, humidity, rainfall, vegetation characteristics, and so on. Additionally, the economic, social, and cultural conditions also play an important role in KA disease propagation. Microgeographical factors and natural conditions influence the prevalence and development of VL (Fig. 3.2). Microgeographical factors acquired from various sources, like ground-based observations and satellite data, are highly useful for disease prevention and have been used to forecast epidemics which is imperative for the preparedness of health systems to cope up with such outbreaks (Nieto et al. 2006; Bhunia 2014; Mandal et al. 2017). Moreover, the linkage between the vector and the natural reservoir turns into a promising factor toward keeping an endemic status for leishmaniasis.

The environmental variables in the biosphere are interdependent and closely associated with the geographical variables. In the tropical and equatorial regions, there is warm and rainy weather favoring the *Phlebotomine* sandfly ecology (Lewis 1971). With the variation of global climate, the geographical distribution of VL in the world appears to be changing, and also the niches of *Phlebotomine* species could be increased. Moreover, people are migrating among the countries due to globalization of economy, thereby increasing the connection of the people with *Phlebotomine* niche where leishmaniasis is either embryonic or absent (Aversi-Ferreira et al. 2015).

Phlebotomine species are mainly observed in forest areas, but are also observed in open and urban areas. The knowledge about leishmaniasis is inadequate; several factors are still unidentified. Due to the climate change and with the impact of global warming, niches of vectors and reservoirs species have been changed.

Fig. 3.1 Transmission of VL (reservoir-vector-host)

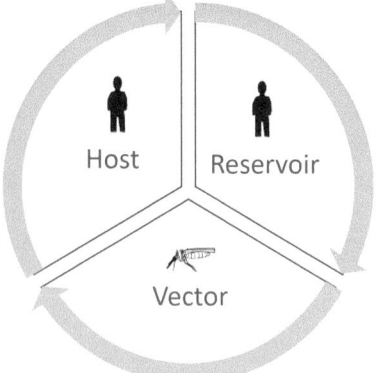

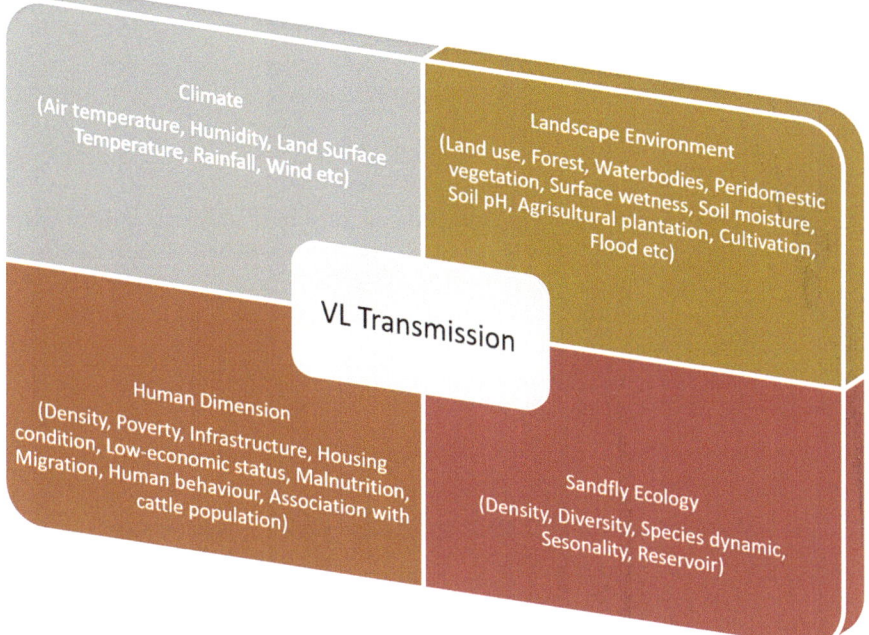

Fig. 3.2 Microgeographical factors influence VL transmission

Government and public should work together to develop exact extermination approaches by emerging specific fortifications with control of *Phlebotomine* by using insecticides and eradicating organic material in the peridomicile areas and eradicating the contaminated reservoir.

3.2 Physical Factors

3.2.1 Temperature and Kala-azar (KA) Transmission

The microclimatic conditions of temperature and humidity are of paramount importance for the transmission of disease. In India, sandflies are most active between June and September, with 0 most abundant when the temperature is 27.5 °C–31.0 °C range (Bhunia et al. 2010a). The temperature difference between areas suitable and unsuitable for sandfly was shown to be ±5 °C, and hence the distributions were very susceptible to temperature variations. Temperature has multiple effects on disease transmission, viz.:

- The optimum temperature for the development of parasites ranges from 23 °C to 28 °C. However, the parasite may develop at a temperature ranging from 16 °C to 32 °C.
- Temperature within this range will generally cause an increase in vector abundance, biting rate, and activity level.
- Higher temperature increases the number of blood meals taken and the number of times eggs are lid by the sandfly.
- The immature stages of *Phlebotomine* sandfly have different tolerances to temperature, with egg stage being more sensitive to higher temperature, while the larval stage is more sensitive to lower temperature.
- Higher temperature also increases the mortality rate of sandfly and may have affected the survival by low parous rate.
- Cold temperature induces mortality or diapauses, slow development rates, or reduced host-seeking activity.

3.2.2   Relative Humidity and Kala-azar (KA) Transmission

Relative humidity also plays a fundamental role in the rate of multiplication of the parasite in sandfly, and directly influences the larval development, gonotrophic cycle, and longevity, as well as the duration of the extrinsic cycle of the *L. donovani* parasite. It may also explain the association between the presence of VL and areas with relative humidities of 66–75% in north-east Gangetic plain (Bhunia et al. 2010a, b). Lesser humidity increases the mortality rate and decreases the parous rate of sandfly. Higher humidity lengthens the life span of the sandfly and enable them to infect more people. The interaction between rainfall, evaporation, and temperature modulates the ambient air humidity which in turn affects the survival and activity of sandfly.

3.2.3   Rainfall and Kala-azar (KA) Transmission

Rainfall is one of the climatic variables that aid in the multiplication of sandfly breeding places and increasing humidity, which improves parasite survival rates. The postmonsoon season is a fertile period for the breeding sites, which are numerous. Studies have established complex relationship between kala-azar and rainfall because humidity is very vital for larval development. According to Lysenko (1971), the breeding success of sandflies is more dependent on the duration of rainfall than the intensity. Thomson et al. (1999) identified a reduction in the infection of kala-azar in Sudan as associated with heavy rainfall. It was found that heavy rainfall may have tends to wash away the sandfly eggs and also destroy the habitats and breeding sites. Similarly, an environmental study using remote sensing found an association between VL incidence and annual rainfall (100–<160 cm) in the Gangetic plain (Bhunia et al. 2010a, b).

3.2.4   Peridomestic Environment and Kala-azar (KA) Transmission

The life cycle of KA/VL is primarily related with the ecological factors in the rural or the peridomicile areas that anchorage the sandfly niches and the reservoir hosts with human habitation (Fig. 3.3). The presence of granaries surrounding the houses is found to be a risk for KA transmission. Bamboo trees (*Bambusa arundinacea*) are the common peridomestic plants found as a risk factor for KA. It has been reported that some plants such as *Amaranthus spinosus*, *Musa sapientum*, and *Croton sparsiflorus* are very rich sources for fructose and thus attract the *Phlebotomine* species (Dhiman and Dinesh 1992). The peridomestic trees also provide gloominess and accordingly generate dark and humid environs around the vicinity of the houses, generating apposite resting places for sandflies. Certainly, the presence of swine species in the peridomicile areas is a significant menace of the propagation, and the transmission has also linked with the cattle population and possible wild reservoirs host of *Leishmania* and feed on them, thus escalating the parasite cycle to the human and the livestock/canine populations.

3.2.5   Soil and Kala-azar (KA) Transmission

The distribution and abundance of sandflies (*Phlebotomus argentipes*) seem to be influenced by structure and composition of soil along with its physical and chemical characteristics. Sandflies require moist soil rich in organic and nitrogenous matter to breed (Singh et al. 2008a). Subsequently, inorganic constituents of the soil were found to affect sandfly breeding. Climatic variables influence soil carbon stocks through their effects on vegetation and through their influence on the rate of decomposition of soil organic matter. In general, hematophagous insects (*P. argentipes*) are attracted by CO_2 concentration gradients, and the attractiveness of other host odors is synergized by CO_2 (Gibson and Torr 1999), because they are dependent on blood for their follicular development and subsequently laying of eggs to maintain their progeny except a few who are autogenous in nature (Ghosh and Bhattacharya 1989).

Soils of the pH from 6.5 to 8.5 have also been identified as one of the major risk factors for VL in several earlier studies in India (Singh et al. 2008a, b). Soil pH enhances its capability of retaining water, and successful growth and abundance of edible shrubs and plants may influence the larval development. Efforts made toward reducing the alkalinity of soil might reduce *P. argentipes* breeding. Subsequently,

Fig. 3.3 Peridomestic environments and kala-azar transmission

soil temperature is also an important factor in chemical reactions and biological interactions in the soil, since it may contribute to maintain moisture conditions in soil/subsoil level, which in turn suits the breeding and propagation of immature stages of sandfly as well as suitable adult resting habitats in the microfocus (Sudhakar et al. 2006). Hence, the properties of soil have played an important role in enhancing the breeding potential of vector sandfly as well as transmission of KA.

3.2.6   Natural Vegetation and Kala-azar (KA) Transmission

Sandflies would not lay eggs arbitrarily and they must make out appropriate sites for larval growth (Killick-Kendrick 1987). The presence of soft stem vegetation indicated as risk factors of VL transmission (Dhiman and Sen 1992). It has been reported that some plants such as *Amaranthus spinosa (Amaranthaccae)*, *Musa sapientum*, and *Croton sparciflorous* are very rich sources of fructose and thus attract *P. argentipes* (Dinesh and Dhima 1991). Ranjan et al. (2005) scrutinized that bamboo trees (*Bambusa arundinacea*) near houses are one of the significant risk factors for VL, because these trees provide shade and consequently fabricate dark and humid atmosphere around the vicinity of the houses, creating suitable resting sites for sandflies (Fig. 3.4). Furthermore, the bamboo trees aid to prop up high fructose-containing climber plants such as *A. spinosa (Amaranthaceae)*. In VL endemic areas in Sudan, *P. orientalis* distribution is associated closely with *B. aegyptiaca* and *A. seyal* tree species (Elnaiem et al. 2002). Ground survey in the VL endemic region of Bihar,

Fig. 3.4 Peridomestic vegetation and kala-azar transmission

India, showed that vegetation sustains the flourishing of the adult population by supplying food and shelter (Sudhakar et al. 2006); and the authors also investigate that peridomestic vegetation within 5 km radius of human settlements in VL endemic foci in favorable summer and monsoon months consists of edible shrubs and plants such as banana, agricultural plantation, orchards, etc., with very soft stem. Colacicco-Mayhugh et al. (2010) suggested that both the species *P. alexandri* and *P. papatasi* are linked with desert, scrub vegetation in Morocco. The soft stem vegetation not only helps to pierce and drink the plant liquid by sandfly species but may also provide as supply of sucrose of infected sandfly species for maturity and reproduction of *Leishmania* spp. within its gut. In an earlier study, Viergever et al. (2005) reported that the paper flower indemnity the *Leishmania* protozoa in the sandfly's gut. He also suggested that the paper flower also enduringly harms the leishmaniasis transmission from sandfly to human or other hosts. Consequently, the lower spatial resolution of satellite data may be counterbalanced by the remuneration of their regularity of observation, minimum cost, and alleviate of accessibility. Moreover, ground-based observations on ecological research are essential to enhance the association that exists between different vegetation characteristics and sandflies.

From spaceborne satellite data, vegetation indices (VIs) were predominantly applied in studying the vector-borne diseases. VIs develop the strong contrast in the reflectance of vegetation in the visible (VIS) and infrared (IR) spectral wavelengths. Multi-temporal satellite data were obtained with different spatial and spectral resolutions, summarizing the photosynthetic capacity or "greenness index" termed the normalized difference vegetation index (NDVI), the most commonly used vegetation index in KA disease transmission modelling. Several VIs were developed to define vegetation vigor using ratios of spectral bands; the most frequently used wavelength were shortwave-infrared (SWIR), near-infrared (NIR), and RED. Several studies have shown a higher concentration of VL cases in less vegetative areas compared to high vegetation density zone (Bhunia et al. 2010a, b; Elnaiem et al. 2003). Thompson et al. (2004) suggested a positive association between the risk of American VL (AVL) for each square and vegetation biomass, but an inverse relationship was also found between the changes in NDVI from the wet to the dry year. In another study in Brazil, an inverse dependency relationship was found between the mean NDVI values in comparison with the incidence of disease in humans at the 95% level of confidence interval (Bavia et al. 2005). Cross et al. (1996) used a combination of mean monthly NDVI data from the NOAA-AVHRR intended for the period between 1982 and 1994 to predict the geographic and seasonal distribution of *Ph. papatasi* in South West Asia and established 0.00 ± 0.06 NDVI limits for vector presence. Deriving the total leaf area in a plant canopy aids the scientist to estimate the amassed and discharged water by an ecosystem, generation of leaf litter, and the process of photosynthesis, which is of further help to the epidemiologist. Elnaiem et al. (2003) suggested that NDVI in the eastern Sudan generally coincide with the sandfly season and should replicate the compactness of trees due to most grasses of the area are highly seasonal and affluent after the start of the rains. Nieto et al. (2006) developed two probability model using Genetic Algorithm Rule-Set Prediction (GARP) and the growing degree day (GDD)-water budget (WB) to envisage the distribution and probable risk of VL in the State of Bahia, Brazil. The author also suggests that the distribution of the VL in the state of Bahia is concentrated in the central Plateau, surrounded by the xerophilous plants. The NDVI measurements from AVHRR analyzed for a 12-year period from 1982 indicated that the range of NDVI that was positively associated with vector presence was 0.0–0.06, and this information was employed to generate a map viewing the areas where sandflies could be there. Bhunia et al. (2010a, b) used NDVI, calculated through moderate resolution imaging spectroradiometer (MODIS) to investigate the correlation between kala-azar positives, and the results showed that most of the cases occurred in nonvegetative areas or low-density vegetation zones. In 2012, Bhunia et al. used minimum, maximum, and mean NDVI values which were computed for each month and compared with the concurrent incidence of kala-azar and the vector density. Like all other environmental parameters, vegetation indices can be applied by health practitioners to understand the spatial variation in the disease incidence and its covariation with the environmental factors. Additional information for different tree species could aid in delimitation of particular tree species, resulting in identification of the vector, as well as disease transmission.

3.2.7 Deforestation and Kala-azar (KA) Transmission

Forest environment can also lead to an increase in the ability of vector to transmit disease. Galati et al. (2003) reported the trap of *Lu. longipalpis* in the jungle environment in the state of Mato Grosso do Sul, Brazil. However, results from another study by Lainson and Rangel (2005) revealed that chunk of *Lu. longipalpis* trapped in the patch of the forest, and the manifested connection of male and females during both dry and wet season, sturdily recommended this to be an important breeding site. Few studies have specifically addressed the impact of deforestation leading to the spread of KA to new areas due to increased ecological instabilities such as deforestation forcing sandflies to move to urban areas (http://www.indiaenvironmentportal.org.in/node/14392). In the rural areas, the extensive deforestation provides the sandfly with a favorable breeding area (Viergever et al. 2005). Mott et al. (1995) suggested that transmission of VL in Sudan arises seasonally in the acacia forest region. He also recommended that deforestation is a major factor in changing the epidemiology and linking the endemic areas of KA. A study on KA noted that the growing figure of VL cases is also partially due to the deforestation (Albuquerque et al. 2009); conversely extensive deforestation increases the risk of human-vector contact (Rotureau et al. 2006). Moreover, the deforestation associated to unplanned urbanization seems to be the cause of peak incidence of leishmaniasis (Mott et al. 1995). Urbanization processes that produce deforested areas make ecological vicissitudes, also alter the forest flora. It also swelling dead plants, thereby growing the volume of decomposed organic material along with the microorganisms on the ground that certainly affect the *Phlebotomine* cycle (Rutledge and Ellenwood 1975). The narrow and adjacent bandwidth properties of hyperspectral data permit for in-depth assessment of earth surface features (Wang et al. 2005), particularly related to the definite plant physiological and structural characteristics (Schlerf et al. 2005), hence obtaining susceptible vegetation spectral indices for their nondestructive assessment to demonstrate the associations between biophysical variables and narrow band vegetation indices more evidently. Correspondingly, the microwave (MW) remote sensing is also essential to delimit the vector or host habitat, allowing to clearly describe the environment nature and structure (Prince 1999). Since, MW remote sensing is able to penetrate through cloud cover and the entire weather conditions at day or night which is crucial for studies relying on a frequently timed origin of the information or for the studies in tropical regions, where regular cloud covers make it difficult to find usable images at the rainy season. It is clear that the technologies we now have to study these diseases are far better than those available to KA epidemiologists in the early years of the last century. The challenge is to make the KA prediction a reliable, acceptable entity in the global community.

3.2.8 Surface Waterbody and Kala-azar (KA) Transmission

Sandfly prefers damp and humid places for laying eggs (Sharma and Singh 2008), and *P. argentipes* required more than 70% humidity for larval development. Bhunia et al. (2011a, b) reported that distribution and spatial extent of surface waterbodies had higher impact of vector abundance vis-à-vis disease propagation. Surface waterbodies influence the VL in two ways: (i) by creating suitable breeding areas by increasing humidity in closeness to waterbodies and (ii) type of surface waterbodies available to the sandfly for oviposition which can alter the insect abundance. Moreover, the areas enclosed by the huge waterbodies and perennial riverbank experience growing dampness of the soil adjacent to the surface, and this can swell saturated low land areas because of flooding during the rainy season, which can reduce the number of larvae. The inland waterbodies, which are extremely predisposed by the hydrological framework in the flat terrain region, might be helpful in creating the environment moister and more humid, eventually encouraging the reproduction of the sandfly populations. Moreover, water availability near saturated grounds, or along perennial riverbanks along with high density of waterbody, allow episodic prosperous during the dry seasons by reaction of lesser volumes of rainfall and can also be effective in extricating sandflies.

3.2.9 Topography and Kala-azar (KA) Transmission

Topography is important as it influences the temperature, humidity, and local environmental conditions and has a propensity of greater variations which affect the dispersal of vector negatively (Bhunia et al. 2010a, b). Spaceborne earth-observation sensors provide new perspectives in digital elevation model (DEM) generation for virtually any location on earth. The Advanced Spaceborne Thermal Emission and Reflection Radiometer (ASTER), Shuttle Radar Topography Mission (SRTM), and Advanced Land Observing Satellite/the Panchromatic Remote-sensing Instrument for Stereo Mapping (ALOS/PRISM) provide unique DEM data with 30–90 m resolution for the entire globe. These models have been used widely in the ecological and environmental sciences for estimating topographic parameters such as absolute and relative relief, slope, aspect, and ruggedness (Petri and Kennie 1990). Bhunia (2014) investigated the degree and extent of relationship between topography-derived variables and VL incidences through Pearson's correlation coefficient. Results showed that higher density of VL incidence has been recorded from the higher absolute relief; it may be due to the lower altitudinal variation. However, these topographic conditions are conducive for water pooling and local depressions. In the flat alluvial plan, microtopographical variation generates waterlogging which contains the surface dampness of the region. Hence, identification of area-specific topographic features has important implications in classification of VL endemicity for specific micro regions. Moreover, fewer slopes and gradients may aid in the prolonged water stagnant condition in Gangetic plain, providing suitable moisture for breeding and propagation of sandfly in the nearby hamlets.

3.2.10   Land Use/Land Cover and Kala-azar (KA) Transmission

Land use/land cover (LULC) variables are directly associated with the presence of sandfly (Feliciangeli et al. 2006). Investigation carried out by Sudhakar et al. (2006) recommends that sandfly populations are mostly prejudiced by three variables, namely, vegetation, waterbodies, and settlements, which sustenance the adult sandfly populations by providing food and shelter. Nevertheless, the comparatively high wetness in areas neighboring nonperennial rivers and other limited waterbodies also provide an imperative role in preserving sandfly density, encouraging transmission of disease (Bhunia et al. 2011a, b). The effect of LULC is usually considered in this connection, but has not previously been analyzed as potential indicator of *P. argentipes* habitats comparing vector suitability vis-à-vis kala-azar incidence. Yet, assessments of suitable climate and vegetation do not only relate to *P. argentipes* ecology but can also indicate availability of definite hosts. Definitely, studies of the effect of remotely sensed land cover structures in VL endemic foci have shown that marshy land, dry and moist fallow tracts, inland surface waterbodies, plantations, and settlements are all strappingly connected with VL incidence (Marzochi et al. 2009). Bhunia et al. (2012b) suggest that agricultural land and paddy fields do not form well-assimilated associations. For example, the propagation of disease earlier happened over extensive areas from where it has now been eradicated by agricultural expansion and insecticide usage. It is recognized that in regard to rice cultivation, modernization is possible to be allied with augmented flooding and limited vector breeding prospects (Müller et al. 2011).

However, the high spatial and spectral resolutions satellite data provide opportunities to explore transmission patterns, distances, and environmental factors (Bhunia 2014). To classify the LULC characteristics, Advanced Very High-Resolution Radiometer (AVHRR) sensor, Moderate Resolution Imaging Spectroradiometer (MODIS), Landsat Thematic Mapper (TM), Enhanced Thematic Mapper+ (ETM+), Advanced Wide Field Sensor (AWiFS), Indian Remote Sensing (IRS)-Linear Imaging Self-Scanning (LISS)-III, and LISS-IV sensor data have been used. However, despite the advancement made, the exactness of most sandfly distribution maps is still not sufficient and comprehensive for planning and execution of control program. Bhunia et al. (2012a, b, c) suggested the use of LULC data at different hierarchical level and identifying the possible LULC classes to be considered for KA distribution as well as suitable habitat of the vector. Hence, LULC variables should be taken into account to investigate the relations between KA control and variations in LULC to observe ecological possessions at the landscape scale and, probably, to explain vindication actions that may remedy any adverse environmental impacts.

3.3 Demographic Factor and Kala-azar (KA) Transmission

Kala-azar (KA) is a poverty-related disease. It affects the poorest of the poor and is associated with malnutrition, displacement, poor housing, illiteracy, gender discrimination, and weakness of the immune system and lack of resources (Desjeux 2001). KA also tends to affect relegated societies, particularly people living in mud houses, near to water resources, and in the locality of accrued rubbish, sewerage, and farms of livestock. Research report stated that inhabitants in underdeveloped houses with mud/brick walls, crack in the wall, thatched wall, and mud floor and lack of personnel protection measures and economically driven migration and employment bring people into contact with infected sandflies (Kesari et al. 2010). The role of domestic animals in proximity of the houses has been emphasized as a risk factor for VL (Bern et al. 2010). However, infected population and low immunity also play an important role in sustaining VL transmission. Desjeux (1996) identified VL patients with the lowest socioeconomic status, having nominal political supremacy to impact the decisions of the government and a very partial capacity to undertake the expenses of the disease. Thakur (2000) investigated that most VL patients had earned less than US$ 1 per day (45 Indian Rupees) that may not be a risk factor, but it can impact to malnutrition, low living status, high illiteracy, poor sanitation, and economic status. Over the last decade, there have been at least seven attempts to estimate the financial and economic costs incurred by VL cases in India (Sundar et al. 2010), Nepal (Adhikari et al. 2009), or Bangladesh (Sharma et al. 2006). The principal cost driver was observed to be the income lost because of illness (i.e., an indirect cost), which may correspond to 60% of the total household expenditure (Meheus et al. 2006). However, VL is still hidden, as cases typically found in isolated areas with inaccessible to services, and caseloads are not documented regularly, with 5–8 unreported cases for each case that is documented (Das et al. 2010).

3.3.1 Housing Characteristics and Kala-azar (KA) Transmission

P. argentipes are endophilic in nature (e.g., breed inside the cowsheds and human dwelling). Kesari et al. (2010) reported that the prevalence of KA and sandfly is high in poor housing condition in India. In human dwelling, sandflies are typically observed inside crack and fissures of the wall and loose bricks, behind furniture, underneath the beds, and in unfilled boxes and hangings in the living room. Research work also reported that mud-plastered wall and cattle shed are suitable for the development of *Phlebotomine* species in ISC (Kesari et al. 2010), Brazil (Badaró et al. 1986), and Kenya (Ryan et al. 2006), respectively. Outdoor resting places for sandfly are reported as bushes, rodent's burrow, rate roles in trees, and bases of banana clumps (Ahammed et al. 2016). Therefore, improving construction of houses with concrete floor and cement plaster wall and cemented/asbestos roof might predominantly lessen the vector population. Kumar et al. (1995) suggested that filling of cracks and crevices in walls with mixture of lime and mud reduced the density of vector population. Moreover, sandfly population is likely to be higher in the cattle shed, as it provides high organic content (e.g., cow dung and urine of the cattle) and loose wooden planks (Fig. 3.5).

3.3.2 *Population Density and Kala-azar (KA) Transmission*

Several hypotheses have been delineated by the qualitative and quantitative assessment that the VL disease may be strongly related to socioeconomic variables (Bhunia 2014). Population density (i.e., population per square km) is positively

Fig. 3.5 Housing characteristics and kala-azar transmission

correlated with the reported incidences during the period. The sandfly vector in India, *P. argentipes*, is endophagic (e.g., biting occurs inside houses) and bites in the evening (WHO 2001). On the other hand, expanding human population living in overcrowded conditions with inadequate housing and sanitary facilities increases the likelihood of VL infection (Mendes et al. 2002). Thus, for example, greater density in terms of household members per room could attract more sandflies, growing an individual family member's exposure to sandflies and, consequently, the possibility of acceptance of an infected bite (Reithinger et al. 2010).

3.3.3 Family Size and Kala-azar (KA) Transmission

Family size is also a significant variable for VL transmission in this region (Pascual Martı'nez et al. 2012). Bhunia (2014) reported that large family size is positively associated with the average annual incidence of VL in India. This seems reasonable because when sandflies are more active, it results in increased chances of sandfly bite and disease transmission. Alternatively, small family size can be attributed to: education of women, aspiration for higher materialistic standard of living, etc. Ranjan et al. (2005) reported that the incidences of VL were higher among cases with the past of KA history among the family members in the past year. The presence of VL cases in the family might aid the transmission of this disease in the presence of sandfly vectors and other conditions favorable for completion of transmission cycle within the house (Bern et al. 2005). Therefore, the areas that experienced large family size could put in place increased preventive measures to help decrease the number of VL cases.

3.3.4 Illiteracy Rate and Kala-azar (KA) Transmission

The distribution of literacy skills in a society is particularly important because inequalities in literacy contribute directly to inequalities in income and occupational status: people with low levels of literacy are restricted in their access to certain labor markets, while those with high levels of literacy are more likely to attain high-paying jobs. The geographical difference of illiteracy rate (both the male and female) was considered as an important socioeconomic indicator for KA transmission (Bhunia 2014). The literacy levels and the economic eminences on the occurrence of KA at the different scales have been depicted by other researchers and scientists (Adhikari et al. 2010). In earlier studies in south Asia, a strong association between kala-azar and poverty has been suggested (Thakur 2000). The results of our analysis illustrated that the areas with unemployed population and higher illiteracy were measured as highly threat areas for transmission of kala-azar. It may be because the individuals do not have acquaintance of the symptoms of the disease and are unconscious of disease prevention techniques (Ross and Wu 1995).

Moreover, female literacy rate is crucial for understanding the impact of VL as females come forward for treatment only at the last stage of the disease, which is mainly due to the social and economic stigma (Singh et al. 2006). Hertzman (1994) suggested that increasing the number of literates could help decrease the disease incidence rate. In general, increased literacy improves health awareness. For instance, knowing about disease features may help an individual in taking timely healthcare decisions (i.e., identifying it at an early stage), or knowing common disease control measures can drastically reduce the risk of contracting the disease. Since, it is best to think of improving the overall literacy rate of the entire population as a means of possibly reducing the VL disease incidence rate.

3.3.5   Unemployment and Kala-azar (KA) Transmission

Moreover, the higher unwaged population is considered as a proxy of income status of the geographical area that replicates a poorer living standard and escorts to an amplified risk for transmission (Bhunia 2014). This is also reasonable since nonworkers spend more time than marginal and main workers at places where vectors are present and, hence, are more likely to contract the disease. This indicates higher the nonworking populations in an area, income level is lower, thereby, live in thatched dwellings with mud floors and provides suitable conditions for KA transmission. Improving purchasing power for adequate economic accessibility may be needed without micronutrients such foods, such as animal flesh, eggs, and milk product, in their diet, in children living poor rural communities cannot be expected to meet their nutrients requirements and the childhood malnutrition rates will remain high. On an average, the household spends $134 on the treatment of family member suffering from kala-azar. This represents more than 50% of what is spent on a family member in an entire year (Thakur et al. 2008).

3.3.6   Urbanization and Kala-azar (KA) Transmission

The periurban areas are the significant risk aspect for VL transmission, as they preserve the places for *Phlebotomine* species and natural reservoir hosts. In urban periphery, poor population is frequently inhabitant as they are forced to live far from downtown, thereby increasing habitat. Such areas are embraced with forests, shrubs, poor housing conditions, and mixed dwelling (e.g., single room used for human and cattle population) and became suitable place to rise contact of sandflies with the people along with domestic reservoir hosts for leishmaniasis. As such, these populations have limited access to basic sanitation and sewerage, therefore generating unhygienic environment for disease transmission. Sometimes, infected people come to urban areas for their treatment and reside in periurban areas due to their limited and scant financial resources. Such city periphery provides suitable condition for the transmission of disease since the contact among vector-hosts-reservoirs is maximized.

3.3.7 Migration and Kala-azar (KA) Transmission

Zoonotic form of leishmaniasis is primarily predisposed by migration from rural to urban areas, deforestation, environmental variations, and dissimilarities in the ecology, or vectors and reservoirs. For KA disease perspective, migration is pertinent in several ways. The key mechanism is that migration leads to new exposure to the infectious agents – either because the migrating inhabitants interchanges into new areas and becomes in touch with infective patients or environments or since the moving population transfers the infections and carries it to hitherto unexposed inhabitants at their point of arrival. Migration due to the pilgrimage (particularly where a huge chunk of people is congregated at one place under primitive sanitary situations) may cause to the blowout of disease. The lower economic class people implied higher prevalence of trachoma (due to proximity to cattle) and dracunculiasis (owing to unsafe water), the leishmaniasis infection comparatively high, owing to the differentials in exposure compared to stable populations (Aagaard-Hansen et al. 2010). Moreover, the stirring populations are often predominantly susceptible to the infective agent to which they are uncovered – either due to nonimmunity or as a consequence of malnutrition. WHO (2002) reported that migration, urbanization, and deforestation have largely contributed to the increase of ZVL in Brazil, Columbia, and Venezuela. Since the last 40 years, 90% ZVL cases increased in Brazil. During this period, massive migration occurred from rural to urban areas (Costa et al. 2005). Temporary migration owing to conflict and natural hazards (flood, droughts, etc.) may influence to the conception of new epidemic foci of VL as nonimmune inhabitants traveled into endemic areas, and on their return, they bring in the parasite into a formerly nonendemic area. Expatriate inhabitants unprotected to loss of assets and succeeding hazard may appear enormously high mortality.

3.4 Case Study 1: In Situ Observation of Geographical Characteristics and Kala-azar Endemicity – Case Study in Mahua Block, Vaishali District (Bihar, India)

3.4.1 Introduction

Environmental systems are complex, and it is possible that modifying factors may be present that are confounding any association that exists between ecology and kala-azar risk. The transmission of kala-azar is dependent on the presence of environmental conditions, which have been linked with the development of both the parasite and vector (Bhunia 2014). The presence of spatial variation in the underlaying environmental risk factors and the estimation of their effects remains a substantial area for further research. In an endemic region of KA, the survival of the sandfly vector appears to depend largely on climate or, at least, temperature, rainfall, and relative humidity (Bhunia 2014). Tobler (1970) stated that "everything is related to everything else, but near things are more related than distant things." The interactions that result in disease are also not exception to this rule. Environmental influences, which are known to be unevenly distributed across space, are likely to have the most influence on disease with in local geographic realms. However, there are different ecological niches for the various species of *Leishmania donovani* that transmit visceral leishmaniasis with some preferring dryer climates (Colacicco-Mayhugh et al. 2010; Peterson and Shaw 2003) while others are prone to areas that are more humid. On the other hand, the environment-KA association is not this simple as there are also many other environmental conditions that modify local ecology. Hence, the present study is undertaken to understand the geographical factors influencing the KA endemicity at microlevel based on in situ observations and geoinformatics.

3.4.2 Study Area

Mahua block is located in the central part of the Vaishali district (Bihar, India), extended between25°41′57.028″N–25°52′16.198″Nlatitudeand85°16′45.275″E–85°29′40.931″E longitude (Fig. 3.6). The soil comprises with ash-gray-silt/silt-clay/clayey-silt and has been found development on the eroded and very gently sloping. The region consists of a thick alluvial and fertile soil. The block receives about 85% of the total rainfall from south-west monsoon, and the relative humidity ranges between 70% and 80% in premonsoon and postmonsoon season. This soil is deficient in phosphoric acid, nitrogen, and humus, but potash and lime are usually present in sufficient quantity. The study area is bounded by small rivulets and nonperennial river in north and south of the block. The block is surrounded by banana, litchi, and mango plantations. Paddy is grown mostly in the clayey soil which is known locally as *mathivari*. Sandy loam, which is known as *balsundari*, is particularly suitable for *rabi* cultivation in the study area.

3.4.3 Materials and Methods

3.4.3.1 Analysis of Cumulative Kala-azar Incidence Rate

For the purpose, the total number of kala-azar incidence report during the period 2013–2015 was calculated. The numerator was the sum of all kala-azar cases recorded during 2013–2015, and the denominator was the average number of populations at risk during this period. And finally the incidence rate was calculated per

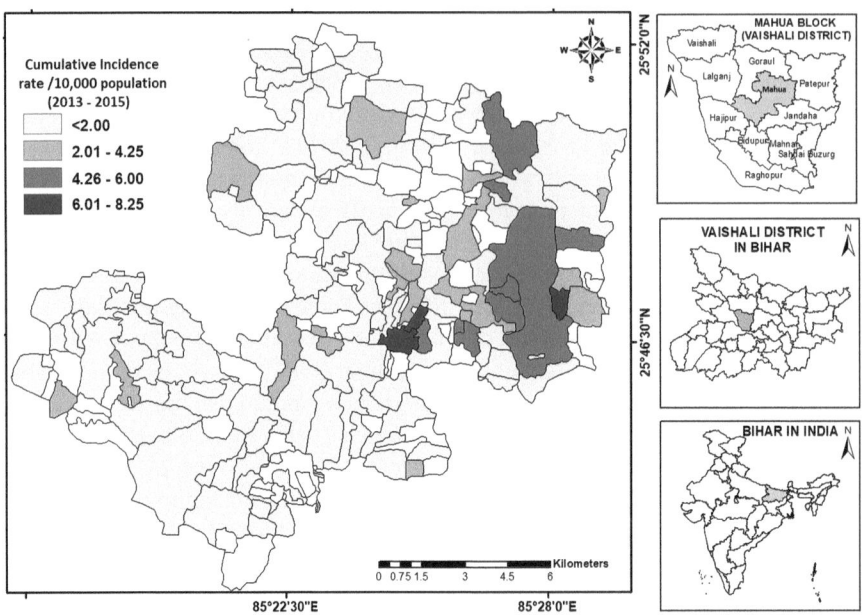

Fig. 3.6   Location map of Mahua block in Vaishali district, Bihar (India)

10,000 populations for each village and integrated into village database of Mahua block.

3.4.3.2   Peridomestic Environmental Characteristics of Kala-azar Endemic Village

A total of 10 villages have been surveyed and total of 10 households have been surveyed randomly based on the KA endemicity. Comprehensive information on peridomestic environment (human dwelling, living conditions distance of agricultural land, types of vegetation and its distance from the households, presence of water bodies and its types, presence of garbage surrounding the households, source of drinking water, etc.) are also collected from KA endemic villages through questionnaires survey.

3.4.3.3   Soil Characteristics of Sandfly Collection Sites

Individual soil samples are collected from the sandfly collection sites of five highly endemic villages. The soils are collected from 5 to 6 cm depth at each site with more than 250 g weight (Kesari et al. 2011). After that, a laboratory test is performed to identify the chemical properties (silicon dioxide (SiO_2), aluminum oxide (Al_2O_3), iron oxide (Fe_2O_3), calcium (Ca), magnesium (Mg), sodium oxide (Na_2O), potassium oxide (K_2O), phosphorous pentoxide (P_2O_5), titanium dioxide (TiO_2) levels, and pH) and moisture content (e.g., ratio between water in the wet soil and water in the dry soil) at each site.

3.4.3.4 Satellite Data Acquisition and Calculation of Vegetation Vigor

To identify the land use/land cover (LULC) characteristics, Landsat8 Operational Land Imager (OLI) sensor data (Date of Pass, November 2, 2015; Path/Row, 141/42) is used. The Normalized Difference Vegetation Index (NDVI) is used to identify the vegetation vigor calculated through near-infrared band minus the red band (ρNIR – ρR) divided by the addition of these values (ρNIR + ρR) (Rouse et al. 1974). The NDVI values range from −1 to +1 where values ≤0 represent water and empty land, while values ≥0 signify vegetation ranging from sparse to green (Krishnaswamy et al. 2009). The minimum, maximum, and mean NDVI value for ten sandflies collection sites are extracted within 500 m buffer zone. The spatial correlation analysis is performed through overlay analysis.

3.4.3.5 Landuse/Land Cover Characteristics and Estimation of Suitability Index

Supervised classification technique is used to classify the LULC characteristics based on maximum likelihood algorithm. Total of ten LULC classes are identified. At the block levels, the buffer zones used for the analysis considered of areas covered by circles with a 500 m diameter. Five villages are selected randomly, having high sandfly density (per trap/per night), and conditional analysis is performed based on probabilistic relationship to understand the role of LULC characteristics in disease propagation and sandfly abundance (Bhunia 2014). The LULC classes are classified as stated by the sandfly habitat suitability, i.e., indicating their potential risk for kala-azar transmission. The scores were derived from the information value (I_j), calculated through the following equation:

$$I_j = \log_{10} \left(\text{class density} / \text{map density} \right)$$

In this analysis, I_j values are ranked with an index ranging from 3 to 0, corresponding to five broad categories of suitability, i.e., very high (>2.50), high (2.00–2.50), moderate (1.00–2.00), low (0.10–1.00), and very low or negligible (negative value).

3.4.3.6 Statistical Analysis

Descriptive statistics are calculated for each soil and peridomestic environmental characteristics. Univariate correlation analysis is performed to establish the association between geographical variables and sandfly density. Linear regression analysis is performed to understand the relationship between sandfly abundance and NDVI value. The statistical significance is defined as <0.05.

3.4.4 Results and Discussion

A total of 100 households have been surveyed. The sandfly is endophagic (e.g., biting occurs inside houses) and pieces in the evening and night (WHO 2010). In the study area, the average household size is calculated as 8.50 ± 3.36. Population living in congested conditions with insufficient housing and sanitary services increases the prospect of KA infection. The mean number of households in the study area is estimated as 1.48 ± 0.74. It is noticed that people affected by kala-azar are extremely poor and live in thatched dwellings with mud floors. The families depend on unreliable sources of income such as casual labor. Subsequently, the results showed cattle shed (mean±S.D. 4.19 m±3.25) and peridomestic vegetation (mean±S.D. 8.42 m±7.27) are very closely associated with KA endemic households. However, all the kala-azar-affected households are far away (mean ± S.D. 693.75 m ± 735.83) from the surface waterbodies. The details of the descriptive characteristics of peridomestic environment are represented in Table 3.1.

The results also showed the average distance between garbage and KA-affected households is 3.95 m ± 2.24. However, the average distance between agricultural land and KA-affected households is calculated as 49.17 m with a standard deviation of ± 35.60.

Chemical properties of soil of kala-azar endemic villages are represented in Table 3.2. In the kala-azar endemic villages, the values of Al_2O_3, Na_2O, K_2O, Ca, and Mg are almost in a similar pattern. The soil pH is slightly alkaline in nature, and the water molecules are ranging between 10.65 and 12.65 (Table 3.2). The sandfly distribution is mainly dependent upon the moist and rich organic content of soil. Sivagnaname and Amalraj (1997) emphasized the importance of soil physiochemical properties in governing sandfly distribution and abundance. In India, Kesari et al. (2000) and Singh et al. (2008a, b) suggested that soil pH and inorganic constituent of soil also influence the sandfly abundance. In the present analysis, moisture content of the soil is high in all KA endemic villages and soil pH also slightly alkaline in nature which is corroborated with the previous research work.

Concerning to the vegetation vigor, the NDVI value of Mahua block is varied between −0.33 and 0.65. The negative NDVI value or the NDVI value <0 is characterized by nonvegetated land (e.g., barren land, waterbodies, built-up area), and the higher positive value of NDVI indicated the dense vegetative area. The cumulative kala-azar incidence rate of each village is overlaid on the NDVI map and the spatial correlation analysis is performed (Fig. 3.7a). Results showed lower positive value of NDVI are closely associated with high KA incidence rate. A 500 m buffer zone is created for the mid-point of endemic village, and the minimum, maximum, and mean NDVI value of each site is calculated. Subsequently, statistical analysis is performed to understand the linear relationship between NDVI values and KA inci-

Table 3.1 Peridomestic environment in and around the kala-azar endemic households

Variables	Mean	Median	SE	SD	COV	Kurtosis	Skewness	1st quartile	3rd quartile
Household size	8.50	8.00	0.49	3.36	39.58	0.00	0.91	5.5	9
Number of rooms per households	1.48	1.00	0.11	0.74	50.26	−0.05	1.21	1	2
Monthly income (in RS.)	2335.42	1000.00	346.13	2398.09	102.68	0.70	1.51	900	3500
Distance between the cattle shed and household (in mts)	4.19	3.16	0.47	3.25	77.62	3.02	1.94	2.32	3.925
Distance between the peridomestic vegetation and household (in mts)	8.42	4.97	1.05	7.27	86.35	−0.05	1.23	3.4	12.195
Distance between water body and household (in mts)	693.75	400.00	106.21	735.83	106.07	−0.58	1.07	100	1350
Distance between garbage and household (in mts)	3.95	3.70	0.32	2.24	56.80	−0.30	0.74	1.975	5.6
Distance between agricultural land and household (in mts)	49.17	50.00	5.14	35.60	72.41	12.63	3.43	30	50
Distance between public health centers and household (in mts)	6.71	6.50	0.33	2.26	33.70	3.37	0.00	6.25	7.45

Table 3.2 Physiochemical characteristics of soil of highly Kala-azar endemic villages

Villages	SiO$_2$	Al$_2$O$_3$	Fe$_2$O$_3$	Ca	Mg	NaO$_2$	K$_2$O	P$_2$O$_5$	TiO$_2$	pH	Water molecules
Dhudhua	41.25 (±1.44)	10.76 (±1.50)	8.13 (±0.96)	5.83 (±0.65)	3.73 (±0.80)	7.98 (±0.67)	2.04 (±0.04)	0.02 (±0.01)	0.01 (±0.011)	7.6 (+0.91)	12.65 (±0.23)
Kajri Bhath	40.25 (±0.82)	12.65 (±1.08)	9.35 (±0.55)	5.23 (±2.53)	2.58 (±0.17)	8.76 (±0.48)	2.11 (±0.18)	0.04 (±0.02)	0.03 (±0.016)	7.9 (±0.77)	12.32 (±0.36)
Soharthi	43.1 (±1.89)	9.68 (±2.07)	8.45 (±0.63)	6.53 (±0.81)	4.12 (±0.68)	8.88 (±1.06)	1.98 (±0.16)	0.06 (±0.028)	0.01 (±0.013)	7.6 (±0.73)	11.69 (±0.72)
Pirpur	40 (±0.02)	11.25 (±1.40)	10.12 (±1.46)	9.87 (±2.19)	3.87 (±0.64)	9.62 (±1.31)	1.75 (±0.18)	0.04 (±0.21)	0.02 (±0.002)	8.9 (±0.63)	11.68 (±0.68)
Gangapur Lachhmi	39.65 (±0.93)	11.46 (±2.45)	8.65 (±0.71)	4.89 (±1.91)	2.59 (±0.67)	10.65 (±0.82)	2.11 (±0.09)	0.02 (±0.22)	0.04 (±0.001)	7.8 (±0.40)	10.65 (±0.74)

dence. Figure 3.7b showed positive correlation with the minimum NDVI value and negative correlation is observed with the maximum NDVI value (Fig. 3.7c). This may be due to the surrounding peridomestic vegetation and soft stem plants that might help sandflies to take sugar meal and resting sites. Therefore, NDVI values have significant role to determine the disease occurrence (Kesari et al. 2011; Bavia et al. 2005). The dense forest surrounding the hamlets controls the temperature and humidity and also retains soil moisture. This circumstance increases the probability of sandfly breeding site, and this condition suggests the possibility of high KA transmission.

Based on the LULC characteristics, the study area has been classified as built-up area, mixed settlement, surface waterbody, dry fallow, scrub land, banana plantation, river, agricultural fallow, crop land, and moist fallow (Table 3.3 and Fig. 3.8). Most of the area in the study site is covered by agricultural fallow. Land use/land cover (LULC) may be particularly important since they are directly linked to the vector presence and KA propagation (Feliciangeli et al. 2006). The importance of LULC of sandfly distribution is emphasized by Bhunia et al. (2012a, b, c) and Fernández et al. (2011) using environmental parameters, including meteorological and altitude data, to estimate the geographic limits for various vector species. Very high suitability for sandfly habitats was attributed to plantation with settlement, banana plantation, moist land, and settlements, but areas associated with agricultural fallow and scurb land also indicated the relatively high suitability.

Spatial variation of KA is mainly influenced by the local environmental variables, namely, vegetation, agricultural plantation, and settlements, which sustenance the vectors by providing food and shelter. Moreover, the high moisture in surrounding environment due to areas of nonperennial rivers and other waterbodies also play a vital role in sustaining sandfly abundance, indorsing the disease propagation at village levels (Bhunia et al. 2011a, b). The presence of scrub land also establishes positive influence for predicting the suitable area of kala-azar vectors (Schlein and Jacobson 1999). The results of the analysis also showed the cultivated fields have negative influence in disease transmission. This may be attributed to destruction of gerbil habitats, local extirpation of reservoir hosts, insecticide treatments, etc. which influence the vector breeding opportunities. However, the agricultural fallow showed moderate suitability of disease transmission. Hence, these mapping efforts may further offer reference information to appraise the influence of kala-azar interventions.

3.4.5   Summary

The environmental features which are particularly important for the delineation of the risk for vector exposure can be captured sufficiently using high resolution satellite data. Moreover, the field-based information can be integrated for predicting risk analysis of KA transmission at microlevel. The present study examines the relationship between geographical factors at microlevel and their suitability for disease

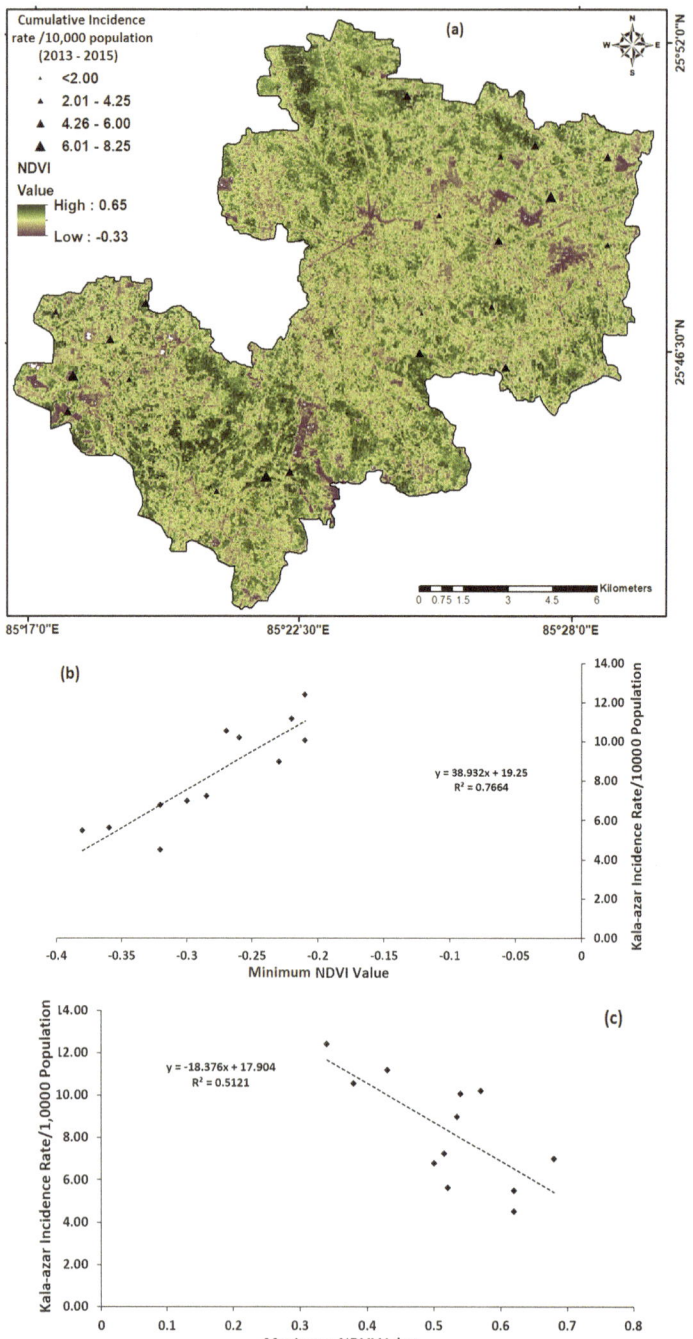

Fig. 3.7 (**a**) Spatial association between Normalized Difference Vegetation Index (NDVI) and Kala-azar incidence rate (2013–2015) of Mahua block in Vaishali district (Bihar, India); NDVI analysis has been derived from Landsat8 OLI sensor data of November 2, 2015. (**b**) Linear regression analysis between minimum NDVI value and cumulative kala-azar incidence rate during the period between 2013 and 2015. (**c**) Linear regression analysis between maximum NDVI value and cumulative kala-azar incidence rate during the period between 2013 and 2015

Table 3.3 Suitability index of land use/land cover characteristics for kala-azar endemicity

Sample villages	Banana plantation	Mixed settlement	Scrub land	Dry fallow	Moist fallow	Crop/cultivated land	Agricultural fallow land	River	Surface waterbody	Built-up area
Dhudhua	0.97	2.95	0.39	−0.12	0.88	−0.14	0.28	0.14	−0.25	0.77
Kajri Bhath	1.52	2.35	0.44	−0.41	1.35	−0.11	0.23	−0.34	−0.26	1.39
Soharthi	1.86	2.61	0.26	−0.43	1.61	−0.19	0.68	0.29	−0.23	0.89
Pirpur	1.51	2.41	0.41	−0.98	1.41	−0.03	0.28	−0.43	0.03	0.69
Gangapur Lachhmi	1.43	2.31	0.38	−0.57	1.31	−0.18	0.61	−0.58	−0.11	1.47

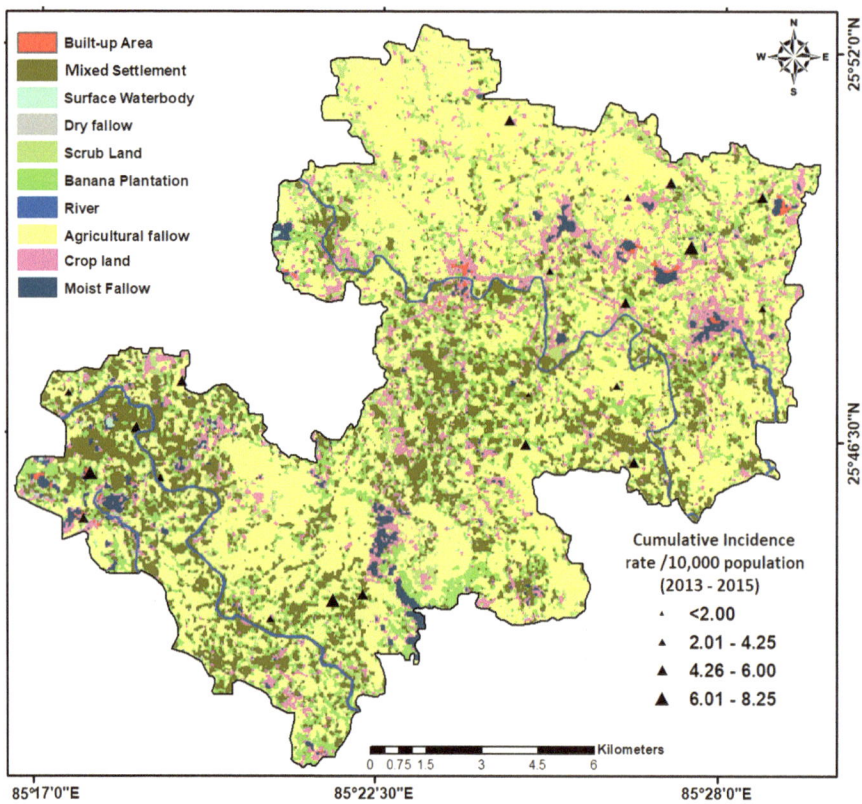

Fig. 3.8 Land use/land cover characteristics of Mahua block, Vaishali district (Bihar, India), derived from Landsat8 OLI sensor

transmission. The results show that peridomestic environment, soil pH, organic content, minimum NDVI, banana plantation, mixed settlement, and moist land are positively influencing the KA transmission. Combined with epidemiological data on disease incidence and microgeographical characteristics derived through free medium-resolution satellite data facilitates the study of the interplay between KA transmission and their environment, which should improve the effectiveness of control strategies.

3.5   Case Study 2: Microtopography and Kala-azar Transmission – Prospects for Control

3.5.1   Introduction

The prevalence of kala-azar/VL varies spatially and temporally due to seasonal changes of weather variables, topography, and associated factors. In lowland plain, drainage is poor, homogeneous distribution of vector breeding habitats, consequently lead to the focal kala-azar transmission and homogeneous human exposure to KA. Sharma et al. (2003) recommended that the uplands have been a refuge from the leishmaniasis vector, and VL infection is infrequent or inattentive. With the increase of altitude, temperature and humidity are also decreased and have a tendency of greater fluctuations affecting the distribution of vector negatively. Bhunia et al. (2010a, b) investigated the relationship between the incidence of kala-azar and topography and reported that high incidence should be expected at low elevation which was found to be the case. Understanding the KA transmission and discrepancies that happen within zones with adjacent immediacy in the relief features would provide the enrichment of an area-specific disease control program for deterrence and epidemic control. Hence, it is a requisite to examine probable factors associated to topography, operating these vicissitudes in spread in an attempt to distinguish susceptible KA-affected villages to assign interferences to be focused at these high-risk areas.

Microtopographical characteristics derived through elevation model would offer a useful base upon which collected data of visceral leishmaniasis (VL) risk could be modelled. As the region is an alluvial plain, the drainage condition is poor, and the presence of poor drainage conditions, marshy land, paleochannels, and scrub land provides suitable sites for vector breeding. Thus, the microtopography comprises little variation of relief features which would provide a rapid method for preliminary stratification of KA epidemic. Hence, the present study is undertaken to identify the spatial association between KA incidence and microtopography.

3.5.2   Study Area

Saraiya block is located in the southern part of Muzaffarpur district, extended between 25°58'8.338"N–26°10'31.633"N latitude and 85°0'52.107"E–85°17'7.256"E longitude (Fig. 3.9). The block has dry and healthy climate with an average rainfall of 1280 mm. This saucer-shaped, low-laying block lies in the great Indo-Gangetic plains of Bihar. The entire district is occupied by alluvial soil. Sandy loam variety predominates east of the Burhi Gandak River. The block is rich in planted vegetation, and its green fields are dotted with groves of mango, litchi, bamboo, and grassland.

3.5.3   Materials and Methods

3.5.3.1   Estimation of Disease Incidence Rate

For this study, 5-year cumulative index (2010–2015) of kala-azar index was calculated, collected from the public health centers (PHCs). The numerator of 5-year annual average incidences was the sum of all KA cases, and the denominator was the average number of people at risk per 10,000 populations during this period 2011–2015.

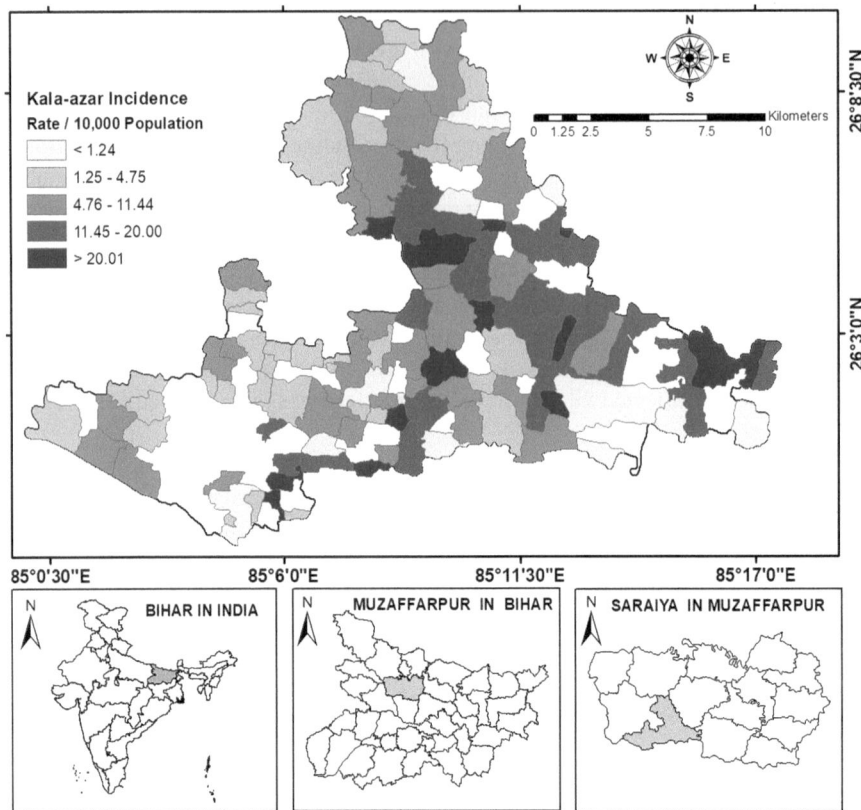

Fig. 3.9 Location map of Saraiya block in Muzaffarpur district, Bihar (India)

3.5.3.2 Conversion of Raster DEM into Mass Points

Geo-referenced Shuttle Radar Topographic Mission (SRTM) Digital Elevation
Model (DEM) data from a 30 m tiles (1 arc second) of the study area were collected
from the Earth Explorer Community (http://srtm.csi.cgiar.org/). The point layer was
generated from the raster DEM in 1 × 1 sq. km interval. The mass points do not
deviate from the input raster than a specified z-tolerance. After that the mass points
are validated with the Google Earth using sufficient point-specified z-units in meters.

3.5.3.3 Triangulated Irregular Network

Topographic modelling uses a triangulated irregular network (TIN) which is a digital representation of the 3-D surface model. TINs are formed by the triangulated set
of vertices which are connected with a series of edges to form a network of trian-
gles. This model uses the Delaunay conforming triangulation method to densify

each segment of the breaklines to produce multiple triangle edges. However, the edges of TINs form contiguous and nonoverlapping triangle facets have been used for estimating topographic parameters.

3.5.3.4 Morphometric Analysis and Spatial Correlation

Morphometric analysis is performed based at 2 m contour interval through grid analysis technique. The various morphometric parameters like absolute relief, relative relief (smith 1935), and dissection index (Dov and Nir 1945) are calculated for each grid. Finally, these are plotted into GIS layer, and radial basis function (RBF) technique is used for calculating smooth surface elevation values over the region. The layer tinting method is applied to classify the progressive elevation zones using geometric interval. Subsequently, the village-wise KA reported cases are overlaid on the map. The number of KA cases is calculated physically to explore the level of influence of disease at each elevation zone.

3.5.4 Results and Discussion

Total of 91 mass points is collected in the study area. The topographic characteristics like maximum elevation, absolute relief, relative relief, and dissection index have been analyzed. The maximum elevation is recorded as 103.43 m in the study area with an average maximum elevation of 63.89 m ± 7.09. The absolute relief in the study area ranges between 43.72 m and 58.86 m. The mean absolute relief in the study area is calculated as 51.08 m with standard deviation of ±3.98 (Table 3.4). The relative relief of the study area ranges from 14.47 m to 62.06 m with a mean and standard deviation of 25.90 m and 10.31 m, respectively. The mean dissection value of the study area is calculated as 0.40 with a standard deviation of ±0.14. The DI value of the study area ranges between 0.22 and 0.84.

3.5.4.1 Maximum Relief and Kala-azar Incidence

The region is flat in nature and the minimum elevation is recorded in the east and south of the block. Due to less altitudinal range, the maximum elevation of the study site is divided into five categories such as (i) less than 62 m, (ii) 63–65 m, (iii) 66–71 m, (iv) 72–83 m, and (v) >84 m (Fig. 3.10). The minimum elevation is found in the central and eastern parts of the block. The highest maximum elevation is found in the south-west of the block. The overlay analysis of kala-azar cases shown in a maximum number of cases are found in the lowest elevated region. The Pearson correlation coefficient is calculated as $r = -0.57$, $P < 0.05$.

Table 3.4 Topographic characteristics of Sariya block of Muzaffarpur district (Bihar, India)

Topographic characteristics	Mean	Median	Standard error	Standard deviation	Kurtosis	Skewness	Minimum	Maximum	95% CI
Maximum elevation (m)	63.89	63.28	0.74	7.09	19.62	3.84	55.96	103.43	1.48
Absolute relief (m)	51.08	51.81	0.42	3.98	−1.10	−0.02	43.72	58.86	0.83
Relative relief (m)	25.90	22.03	1.08	10.31	3.52	1.96	14.47	62.06	2.15
Dissection index	0.40	0.36	0.01	0.14	1.73	1.48	0.22	0.84	0.03

3.5.4.2   Absolute Relief and Kala-azar Incidence

Absolute relief is a maximum altitude of an area that influences the occurrence of KA distribution. The highest absolute relief is observed in the south-east and northern part of the Saraiya block. Moreover, the lower absolute value is recorded in the east, south-east, and small pockets of southern part of the block (Fig. 3.11). In the study site, lower absolute relief has been found suitable for KA propagation. A quantitative analysis of the distribution of VL in relation to absolute relief category divulges a very strong inverse correlation ($r = -0.50$; $P < 0.05$). The maximum KA incidences are recorded in the elevation values less than 50 m.

3.5.4.3   Relative Relief and Kala-azar Incidence

The relative relief (RR) represents the actual variation of altitude in a unit area with respect to its local base level. The maximum number of KA cases is distributed with an elevation range between 25 m and 38 m; however, such significant association was exemplified with the low relative relief region (Fig. 3.12). It reveals a negative correlation ($r = -0.40$) between RR and VL distribution. However, majority of the study area are covered with RR ranges from 25 m to 31 m. Moreover, the high RR indicates more ruggedness of terrain condition and low concentration of VL (Bhunia 2014). Here, despite the low relative relief, the area may be attributed to waterlogging condition which aid to surface dampness in the region. Identification of area-specific topographic features has important implications in the classification of VL endemic for specific micro regions.

3.5.4.4   Dissection Index and Kala-azar Incidence

Dissection index (DI) is an important morphometric indicator which determines the nature and magnitude of terrain condition. The DI value ranges from zero (complete absence of dissection) to one (vertical cliff). Deen (1982) stated that high value of DI indicated high RR and the slope is more. Based on the DI value, the study area is divided into (i) <0.37, (ii) 0.38 – 0.45, (iii) 0.46 – 0.54, (iv) 0.55–0.65, and (v) >0.66 (Fig. 3.13). The highest number of KA cases is recorded from the high value of DI. The lower DI is found in the north and south-west of the block. Moreover, the high value of DI is observed in the central and south-east of the block. A quantitative analysis of the distribution of VL cases in relation to DI category reveals a strong and positive correlation ($r = 0.50$). However, these topographic conditions are conducive for water pooling and local depression with plain slopes. These locations are normally spaces with sandy loam to clay soils where water will accumulate but because of the flat slopes do not drain and retain the moisture and dampness in the neighboring areas.

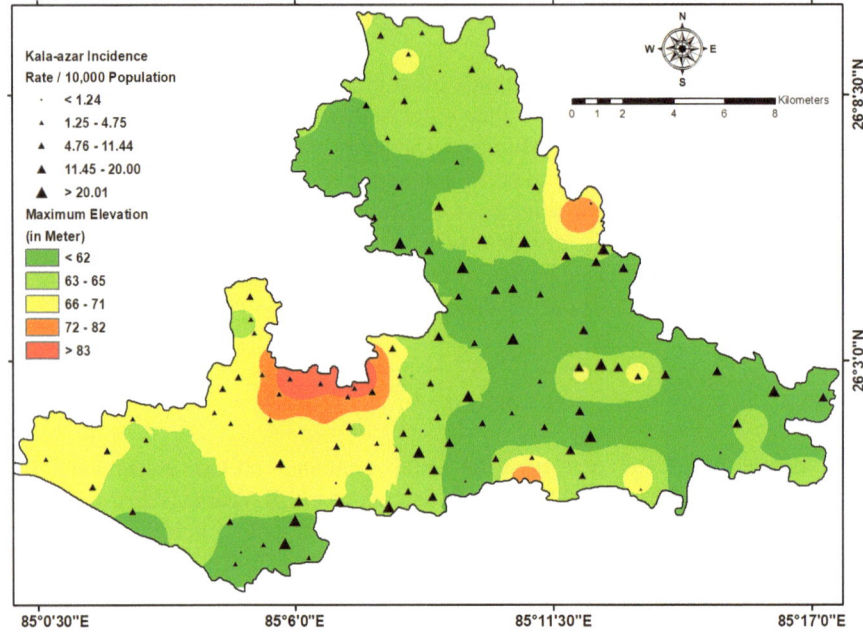

Fig. 3.10 Spatial association between maximum elevation and kala-azar incidences in Saraiya Block, Muzaffarpur district

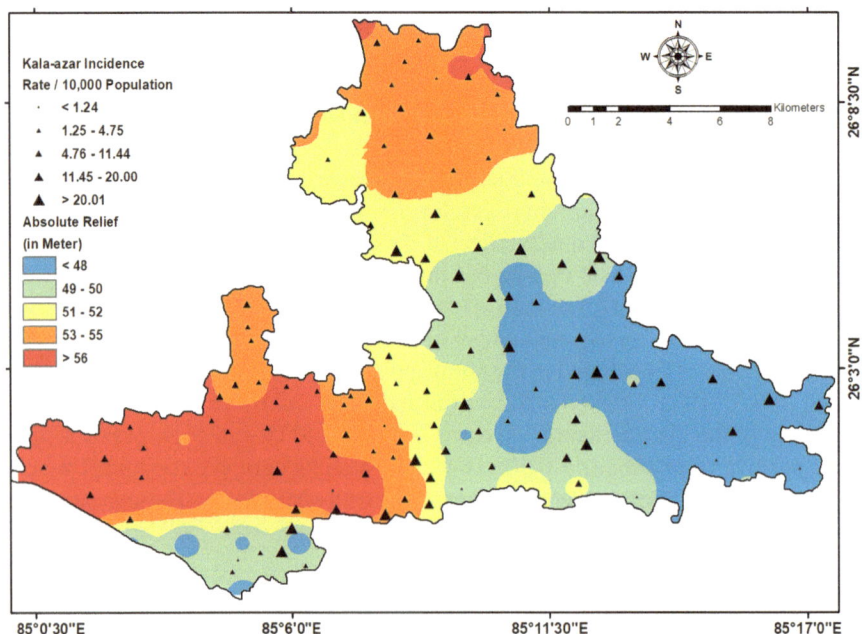

Fig. 3.11 Spatial association between absolute elevation and kala-azar incidences in Saraiya Block, Muzaffarpur district

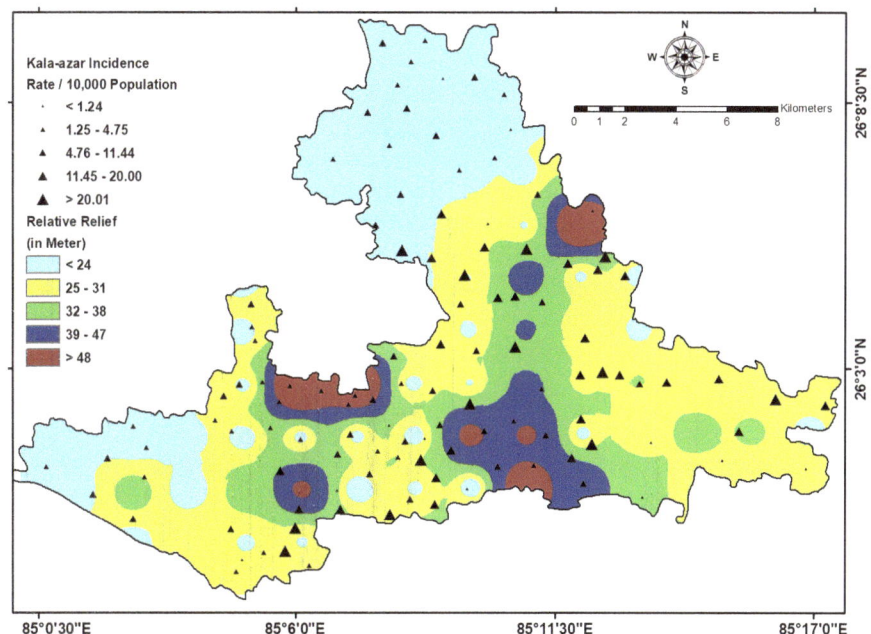

Fig. 3.12 Spatial association between relative relief and Kala-azar incidences in Saraiya Block, Muzaffarpur district

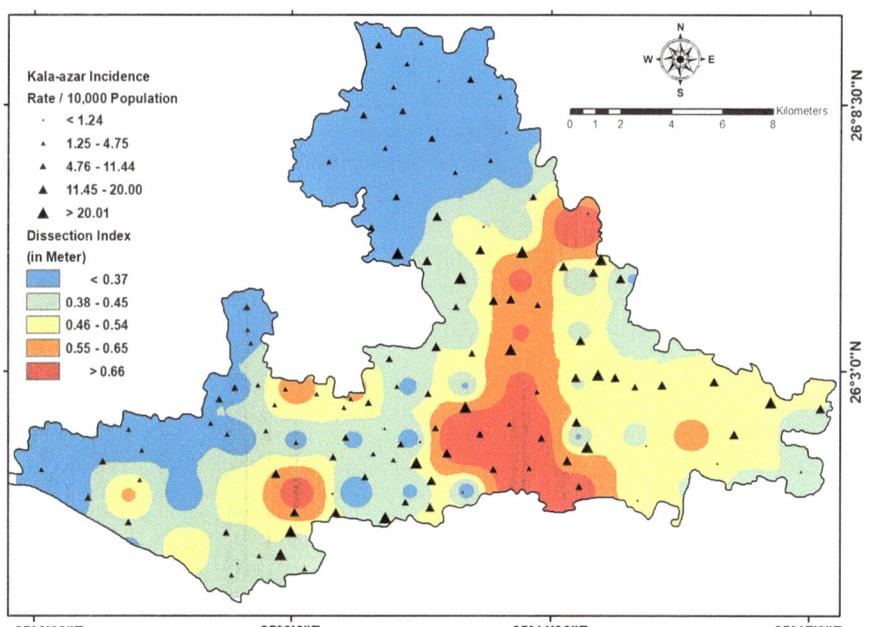

Fig. 3.13 Spatial association between dissection index and kala-azar incidences in Saraiya Block, Muzaffarpur district

3.5.5 Summary

Micro relief is the term applied to minor surface undulations and irregularities of the land surface. The plain areas with gentle slopes are usually characterized by high frequencies of VL incidences, and preferred for habitation of sandfly breeding. Geomorphologically, the region is flat alluvial plain, and the unevenness of land is very less across the study sites. It guaranteed the presence of thick soil in the local area. Such topography rules out the deposition of very coarse sediment by Gandak river. The result showed the high VL occurrence areas were matched with the medium to high dissection index region. These factors may assist to the health planners for predicting and classifying high-risk foci that are highly suitable for KA transmission.

3.6 Conclusion

Kala-azar (VL) is an intricate multisystematic disease and hence needs multidisciplinary team working with the public health agencies, health professionals, and research scientist to attain the optimistic results on the way to its annihilation. Majority of the epidemic outbreak are linked with the natural environment and man-made factors. The local causative factors are required to identify *Phlebotomine* niches and presence of natural reservoir hosts of VL. Thus, investigation of geographical region provided to establish the relationship between environmental variables, favoring the presence of vector and hosts. However, the regional geographical conditions of VL incidence must be analyzed and verified by the public health agencies and health planners. Moreover, the microgeographical characteristics should be castoff to investigate interfaces between VL control and changes in landscape to monitor ecological possessions at the various scale and, perhaps, to express extenuation measures that may cure any adverse environmental impacts. Therefore, geo-environmental aspect may be of value to health interventions and local authorities, who may use this information to regulate and design VL surveillance undertakings as well as to necessitate extenuation programs in the disease-identified areas.

Chapter 4
Open Source GIS and Kala-azar Transmission

Abstract This chapter describes about the Open Source GIS (OSG) software in disease transmission mapping and modelling. The list of OSG software has also been analyzed for mapping and modelling of disease. Current OSG can aid from assimilating different types of maps and charts to epitomize health phenomenon for efficient insistences. Majority of the OSG use choropleth maps and other thematic maps to display spatial patterns of health phenomenon. This chapter also describes the list of open source public health data and sources of specific applications. Few examples of spatial mapping and visualization techniques of public health data have also been illustrated based on the OSG.

Keywords Open Source GIS · Spatial analysis · Data visualization

4.1 Introduction

The initiation of digital mapping and GIS has fully reformed the way human think about and interrelate with the world around them. In 1960, *Ian McHarg* (a landscape architect) first introduced the concept of overlapping of various layers for decision-making. At the same time, *Roger Tomilson* used computing methods and geospatial data for his doctoral research. In 1978, the concept of Open Source GIS can be traced back to the US Department of the Interior. GIS has been used extensively in public health system and is indissolubly related to an exact location in the form of street addresses, postal codes, or geographic locations. With the development of World Wide Web (WWW), GIS data and functionalities have become increasingly available online, resulting in the emergence of Open Source GIS. In comparison to Commercial GIS software, Open Source GIS (OSG) software permits end users at levels to concurrently view the similar up-to-date health data. Open source projects are typically based on programming languages (e.g., "C," Java, .NET), worked on by a community of volunteer programmers. Another advantage of the OSG is to lessen the cost of assimilating GIS into public health practice by offering online training lessons. The progression of OSG since the past four decades has led to its many ground-breaking and impactful applications today.

G. S. Bhunia, P. K. Shit, *Spatial Mapping and Modelling for Kala-azar Disease*,
SpringerBriefs in Medical Earth Sciences, https://doi.org/10.1007/978-3-030-41227-2_4

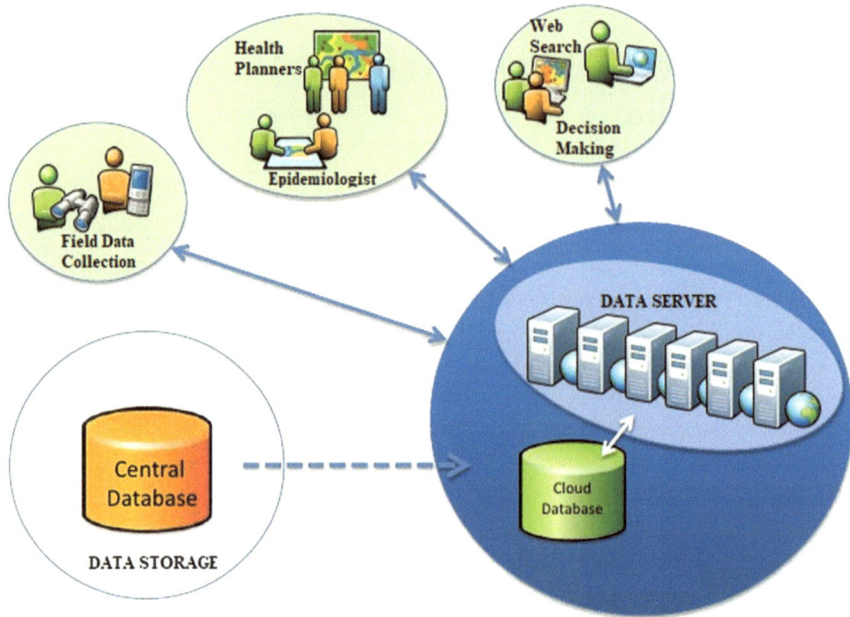

Fig. 4.1 Open Source GIS in kala-azar control program

A good OSG develops from various backgrounds and computer languages; most are not technical, cartographic, or statistical expert (Luan and Law 2014). Per se, the end users are key stakeholders and their requirements and views concerning user interface design are of the principal importance. Moreover, inappropriate map customization selections risk misleading end users, creating suitable customization sceneries strategy as necessity. Current OSG can help from integrating various types of maps and charts to epitomize health phenomenon for systematic tenacities. Subsequently, the animated maps are being underused in prevailing OSG likened with static and interactive maps. Majority of the OSG use choropleth maps, and other types of thematic maps, like graduated symbol map, isopleth maps (e.g., having greater ability to display spatial patterns and represent health phenomenon with a continuous surface), and cartograms, can provide an abundance of additional information. Similarly, beyond the traditional line, bar or pie charts, and unconventional chart types (pyramid chart) commonly used in other domains (Fig. 4.1).

4.2   History of Open Source GIS Software

In 1996, the *University of Leeds* has established a GeoTools project to create a Java-based GIS library for the manipulation of geospatial data. In 2000, Geospatial Data Abstraction Library (GDAL) was set up for GIS applications to sustenance the varying data formats that occur throughout GIS world. It supported 50 raster and 20 vector data formats. In 2006, Open Source Geospatial (OSGeo) software was launched and has been designed to support the collaborative development. In 2011, "Geo for All" was originated with the goal of creating geospatial education and opportunities available to everybody. The software also used supporting applications like Google Earth, Geographic Resource Analysis Support System (GRASS), QGIS, the Feature Manipulation Engine, etc. Since the last decade, there is speedy evolution of using OSG in public health surveillance (PHS). Despite the fast commitment, important disparities occur in the expansion and concert of WebGIS-based PHS among the countries.

4.3   Health Data Mining and Open Source Software

Open source GIS (OSG) software clasps great potential, as it permits for improved teamwork, the distribution of valuable data, and admittance to key resources. OSG provides similar functionality allowing users to interact with the data through a browser from any location or device. With its numerous ecological, government, public safety, and health applications, OSG allows the potential to change the ecosphere. In 2001, Refractions Research has developed an open source program PostGIS to spatially enabled data stored in Postgres database using Java applications. Both PostGIS and GeoServer were incredibly successful projects and are widely used as OSG database and GIS server. Table 4.1 illustrated the list of open

Table 4.1 List of open source software and data for public health mapping and applications

Open source software	Freeware GIS	Specific applications
GRASS (http://grass.itc.it/)	Accuglobe (http://www.ddti.net/)	GeoDa (https://www.geoda.uiuc.edu)
QGIS (http://www.qgis.org)	Google earth (http://earth.google.com)	SaTScan (http://www.satscan.org)
uDig (http://udig.refractions.net)	Diva (http://www.diva-gis.org/)	EpiInfo (http://www.cdc.gov/epiinfo)
MapWindow (http://www.mapwindow.com/)	ArcGIS explorer (http://www.esri.com)	Space time analysis of regional system (STARS) (http://regionalanalysislab.org/)
SAGA (http://www.saga-gis.uni-goettingen.de)		Batchgeocode (http://www.batchgeocode.com)
gvSig (http://www.gvsig.gva.es/)		Spatial libraries in R (http://www.r-project.org/)
MapGuide OS (http://mapguide.osgeo.org/)		
FWTools (http://fwtools.maptools.org/)		

source GIS software platform for public health data analysis. The detailed examples and application of the open source GIS software are described in the Table 4.1.

GRASS GIS*GRASS* stands for Geographic Resources Analysis Support System and is a tool for land management and environmental planning. GRASS GIS is used extensively in academic and government stakeholders (NASA, NOAA, and USGS) which can be qualified to its amicable and in-built GUI. This software is used for image processing, data management, vector analysis, geocoding spatial modelling, graphic production, and data visualization. The software comprises the advantage of 350 robust vector and raster manipulation tools and intuitive graphic user interface (GUI) and reliability.

Quantum GIS (QGIS) is widely regarded as premiere open source desktop GIS, released in 2002. The software integrates analytical functions from GRASS, along with Geospatial Data Abstraction Library (GDAL) data format – into a user-friendly desktop application for execution data apprehending, editing, cartography, overlaying, spatial analysis, manage and export data. QGIS is able to consume data in all formats and can export data into other GIS programs. In February 2018, QGIS 3 brings a whole new set of cartography, 3D analysis tools which boost the mapping software into a state of epicenes. It is a user-friendly interface and has multiple plugins and tools that can be customized. The latest version of QGIS is supported on Mac, Linux, and Windows operating systems.

gvSIG desktop is an open source GIS software, first launched in 2004. The software supported several raster and vector file formats, databases and connect to remote services. gvSIG allows users to create 3D visualizations of their geospatial data out of the traditional 2D views. It includes a suite of CAD tools for tracing geometries, editing vertices, and snapping and splitting lines and polygons. gvSIG provides mobile to record information of rainfall, road networks, etc. It is also supported on Mac, Linux, Windows, and Ubuntu operating system.

OPEN JUMP *OpenJump* software is capable of handling large sets of data, written in JAVA. It can read and write shapefiles along with spatial database and construe numerous vector formats and spatial analysis like buffers, overlays, and vector data. Most important advantage of the OpenJump software is the ability to edit geometry. The software has also the advantages of creating pie charts, plotting, and choropleth maps.

uDig (User-Friendly Desktop Internet GIS) software is most suitable for basic mapping and provide a user-friendly framework to build complex analytic data. uDig is built up with drag and drop interface, editing tools, vector operations, and import base maps. Mac OS functionality makes it robust and powerful, and the software can also work as an independent application or an extension with RCP plugins.

Diva GIS is a free GIS software package for mapping and analyzing data. The software delivers useful free GIS data based on the mapping purpose, for example, to extract climate data for all location on the land.

SAGA GIS is used for automated geoscientific analysis. It allows envisaging along with managing geographic data with the help of maps, graphs, and histograms.

CDC's Epi Info: Epi Info is a free epidemiologic software package on a Windows operating system, intended for the global public health community of practice and investigation. The software has been designed to perform data collection, database creation, data analysis, visualization of data, etc. The software is also intended to vector surveillance mobile applications that vector control staff can use to collect and send data to an analysis dashboard. Epi Info also includes a GIS mapping feature called Epi Map that displays geographic data in choropleth or dot density maps (Table 4.2).

4.4   Mapping and Visualization

Mapping and visualization technology have appeared as a method of exploratory cartography which can support to elucidate, investigate, and communicate distribution pattern. There is a close relationship between visualization and cartography (e.g., art and science of map making), which assists to understand the spatial relationship with the geographical and non-geographical aspects. *Geographic visualization* or *geovisualization* emphasizes on the visualization of geographical data,

Table 4.2 List of data sources for public health mapping and applications

Vector datasets	Attribute files (international-national-local)	Raster datasets
Census (http://www.census.gov/geo/)	WHO (http://www.who.int/research/en)	CIESIN/SEDAC (http://sedac.ciesin.org)
ESRI (http://www.esri.com/data)	WB (http://devdata.worldbank.org/hnpstats)	USGS earth explorer community (https://earthexplorer.usgs.gov/)
EpiInfo (http://www.cdc.gov/epiinfo)	NVBDCP (https://nvbdcp.gov.in/)	Open street map (https://www.openstreetmap.org/)
HealthMap (http://www.healthmap.org)	DHHS geospatial warehouse (http://www.geodata.gov; http://www.nationalatlas.gov)	Natural earth (http://www.naturalearthdata.com/)
EpiScanGIS (http://www.episcangis.org)	GECensus (http://gecensus.stanford.edu)	

which can be used to all phases of spatial analysis in problem-solving process, from expansion of preliminary hypothesis, through knowledge discovery, investigation, demonstration, and assessment (Gahegan 2000). In epidemiological study, *geovisualization* is considered to comprehend not only the development of theory, tools, and approaches for the conception of spatial data; it also comprises understanding how the tools and methods are employed for hypothesis design, pattern documentation, knowledge building, and the enablement in decision-making. The focus in data mining is on detecting the unknown. The broader process of knowledge construction may use the outcomes of mining with more formed data questions. Abduction is flexible as it is not delimited to use existing knowledge structures that aid to elucidate the data offered. In preliminary exploratory phase, abduction process is highly appropriate, concerning probable structures in the data. Data on disease distribution tend to be heterogenous, multifaceted interdependent, indirectly analogous, and interrelated in ways that are not instantly seeming. Due to accessibility in several robust, computerized approaches are developed by machine learning process. Induction becomes a more beneficial and dependable means of knowledge construction. When objects and categories are already well defined, deduction can be useful and typically forms the basis of most inferential analysis (Fig. 4.2). Explorative visualization is an effort to enumerate the correlations, uncertainties, and hidden information.

4.4.1   State of the Art

The present state of art for kala-azar data analysis can be divided into three distinct areas:

(i) Direct visual methods to data exploration and data mining.
(ii) The sustenance of knowledge building and geocomputation through cooperating and collective means.

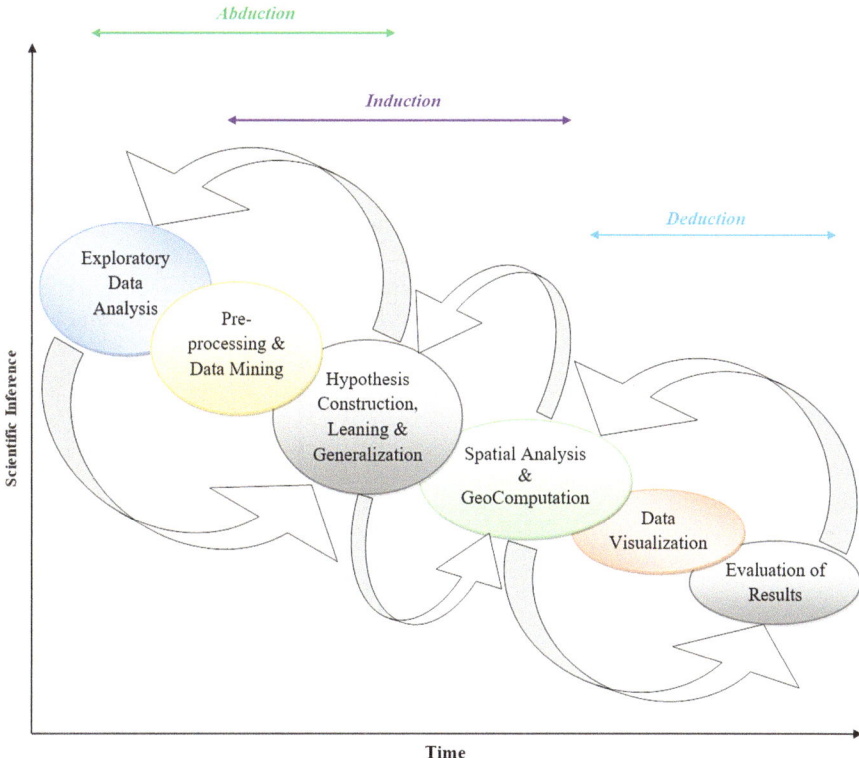

Fig. 4.2 Various phases of scientific method and inferences for visualization of disease data

(iii) The choices of database, data models, and united knowledge construction environments can have on the outcomes of knowledge construction process.

4.4.2 Visualization Approaches

The support of knowledge construction and geocomputation facilitating data exploration task and improving machine learning approaches, *geovisualization* play an important role as an intermediary between human and the computer. Numerous types of graphic data mining and exploratory analysis techniques can be acknowledged. The Exploratory Spatial Data Analysis (ESDA) describes the application of exploratory techniques to capitalize on the pattern appreciation and hypothesis generation capacities of humanoid experts. It creates an opportunity for human and machines to work together in constructing and evaluating the objects and the relationships required by the analysis phase, making the best use of abilities of each. This also provides an environment within which expert scientist can collaborate on complex modelling and analysis activities. However, both these processes help in process-pattern tracking (e.g., visual representation) and processing steering (e.g., interactive environment). Moreover, geographers and statistician are also involved in exploratory analysis which is either chart based or mapped based.

- *Map-based techniques:* This allows to change the visual appearance interactively.
- *Chart-based techniques*: This allows to plot data on a chart or graph. For example, scatter plots use a simple 2D or 3D graph with dot or sphere the position of individual data items.
- *Projection techniques*: This permits statistical performance such as principle component analysis (PCA) and multidimensional scaling to project structure or trend from the data.
- *Iconographic techniques*: This is based on complex symbols such as bar graph, pie graph, stick symbols, etc.

Example
Visualization of spatial distribution cumulative incidences of Kala-azar cases during the period between 2010 and 2015 has close links with population density. This map emphasized the use of static map designed for population density and cumulative incidence of KA patterns revealed and relations understood.

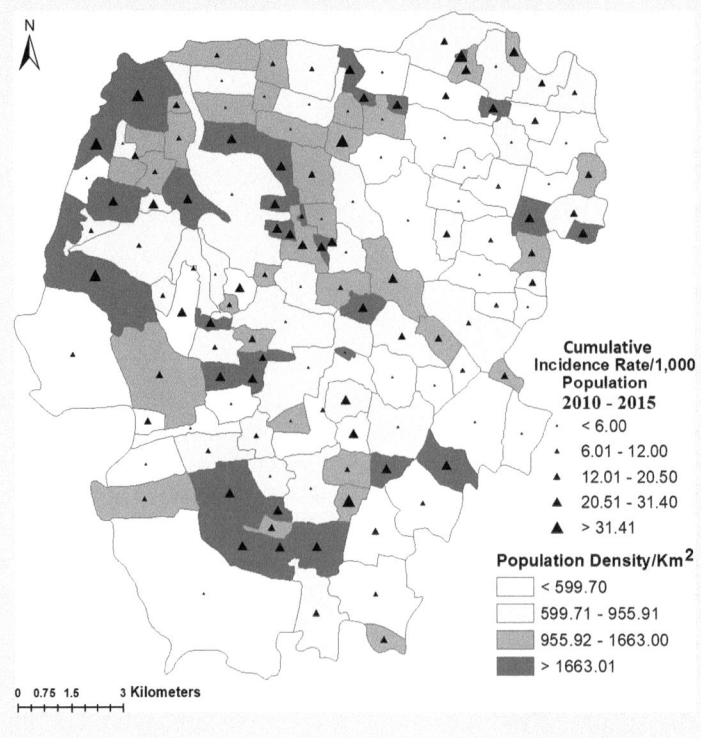

Above figure of Sahebganj block in Muzaffarpur district (Bihar, India) representing the level of concerns, ranging from low (white) to high (dark gray) based on combination of population density that distribute a KA risk evenly over a risk in fact is not homogenous to such an extreme degree that it follows the choropleth boundaries, prepared by QGIS software.

4.5 Spatial/Temporal Analysis of Kala-azar Disease Data

Knowing the distribution pattern of disease consents health practitioners to comprehend beneath mechanism of disease expansion over time and driving aspects of those variations. Comprehensively, a known pattern of the disease distribution might reveal the process of underlaying spatial distribution of disease and association with the physical, environmental, and social conditions (Kienberger and Hagenlocher 2014). Epidemiological data can be represented as clustered, random, and regular pattern, and the analysis methods are grouped into "clustering analysis" or "cluster detection analysis." The georeferenced data along with attribute describes the characteristics of disease locations. Using these data, visualization, exploration, and modelling are conducted to assist in decision building (Fig. 4.3). However, the use of spatial information becomes popular with the integration of GIS and statistics package in processing health data.

The documentation and evidence in heterogeneity in disease incidence across a range provide a geographical range and possibility for targeting preclusion and treatment interferences at high-prevalence areas. Geoinformatics have looked as a new generation of information systems with the competence to direct spatial extents together with public and other magnitudes of interest. These spatial allocations or patterns are used to perceive and enumerate the patterns of disease allotment that may proffer an imminent into a disease epidemiology in health research (Srividya et al. 2002). Figure 4.4 illustrated an example of the spatiotemporal variation of disease distribution in Bihar (India) using free source software.

4.6 Modelling of Kala-azar Disease Data

The incidence of KA is known to be connected to several factors comprising environmental, social, economic, immunological, health services, etc. This section abridges the theory, principles, and approaches of risk assessment epidemiology for investigating exposure-disease relationships based on measure of exposure and measure of disease occurrence. A common measure of disease incidence considered in KA epidemiology is the incidence rate (IR), referred to the new emerging cases.

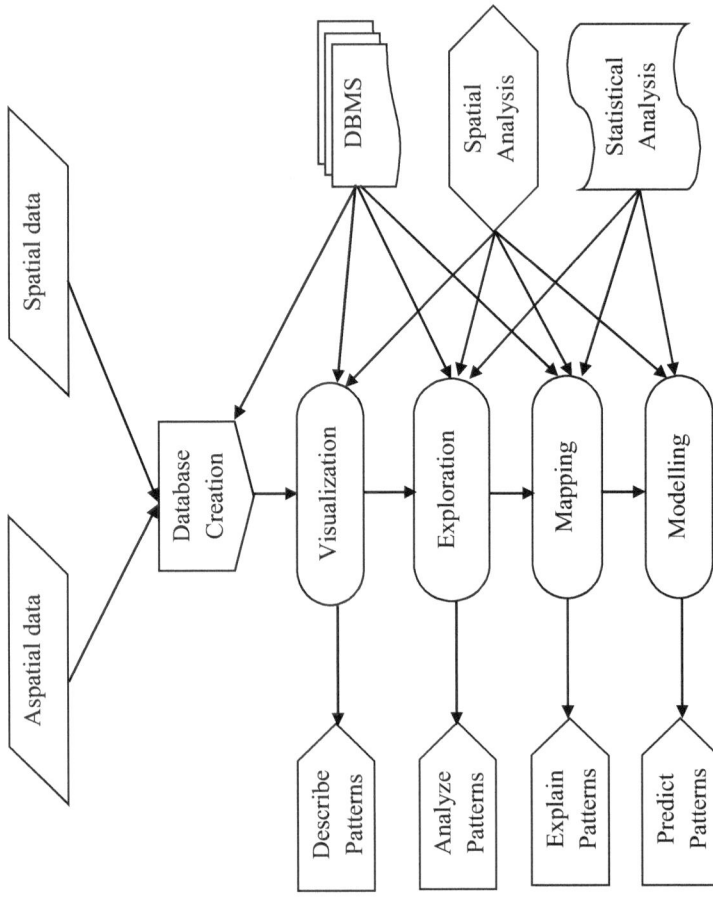

Fig. 4.3 Conceptual framework for epidemiological data analysis through open source software

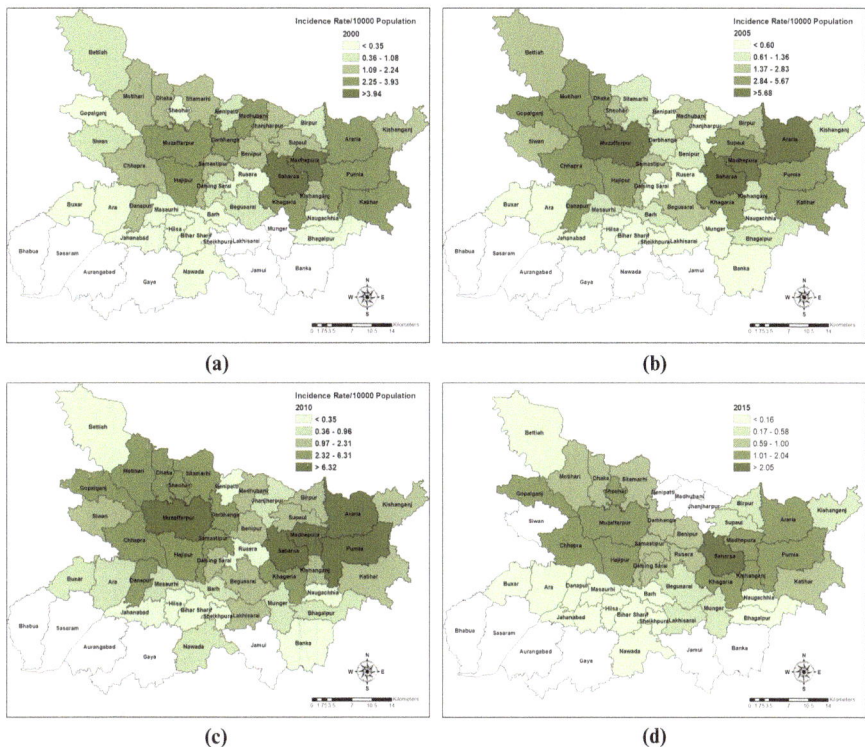

Fig. 4.4 Spatiotemporal analysis of kala-azar incidence rate/10,000 population in Bihar (India), (**a**) 2000, (**b**) 2005, (**c**) 2010, and (**d**) 2015; produced in OpenJump software

4.6.1 Estimation of Incidence Rate

The incidence rate (IR) of kala-azar can be assessed by segregating the following period into intervals of lengths L_j having midpoints t_j for $j = 1...,J$, and approximating a rate for each interval. Let n_j denote the number of individuals who are disease free and still under observation at time t_j, and d_j the number of new diagnoses during the jth interval. An estimate of the IR at time t_j is obtained by dividing d_j by the product of n_j and L_j.

$$\lambda\left(t_j\right) = \frac{d_j}{n_j L_j}$$

The denominator $\lambda(t_j)$ is an approximation to the sum of the observation times on the n_j population members in the jth interval. However, the incidence rate is time dependent based on its origin and the length of time interval. Figure 4.5 shows the KA incidence rate per 10,000 population of Patepur block in Bihar (India) in 2011.

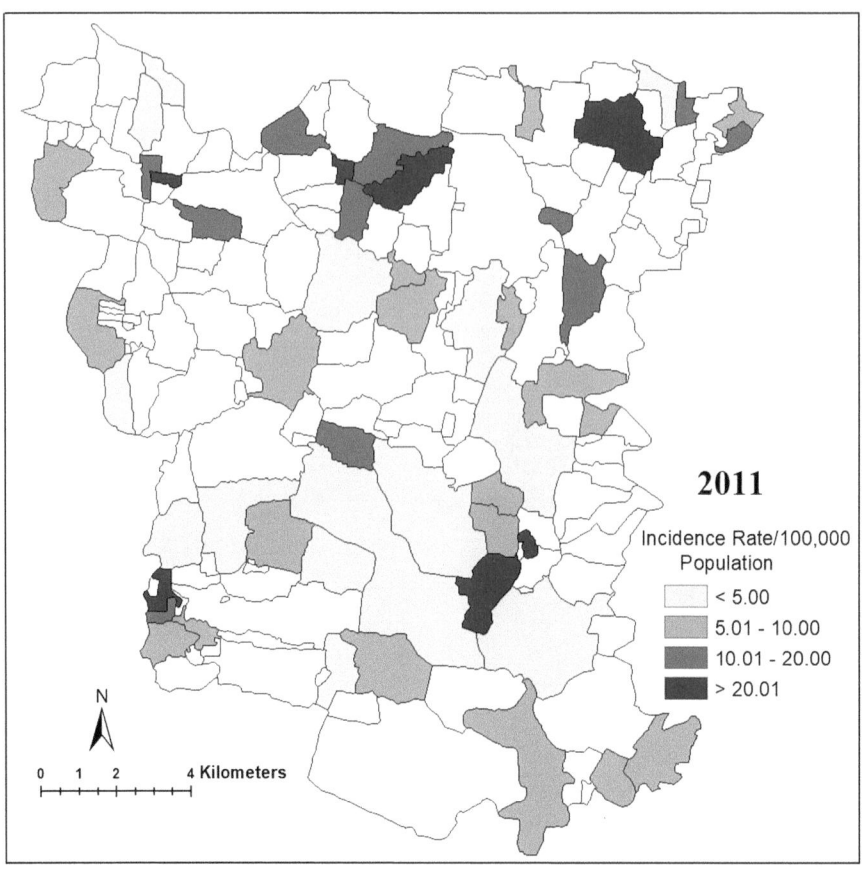

Fig. 4.5 Incidence rate of Patepur block (Bihar, India) prepared by QGIS software

4.6.2 Spatial Association Between Environmental Variables

A spatial association analysis is performed in open source GIS software. The Normalized Difference Vegetation Index (NDVI) analysis has been carried out through raster calculator of spatial analysis tool of QGIS software v2.4.0. After that, *point layer* is created of sandfly trapping site, and the *attribute* (e.g., sandfly density at each point location) is integrated through *spatial join* tool. Based on the NDVI value, the study area is divided into low density, medium density, and high vegetation density zone. *Overlay analysis* is performed to understand the vegetation characteristics and KA incidence (Fig. 4.6). Vegetation vigor has been used for risk assessment, associated with the potential distribution of vector, as it influences the meteorological and environmental variables in its surrounding. The remotely sensed normalized difference vegetation index is used to identify the spatial association between NDVI and *P. argentipes* distribution. The analysis showed low density vegetation cover is closely associated with the high density of the vector.

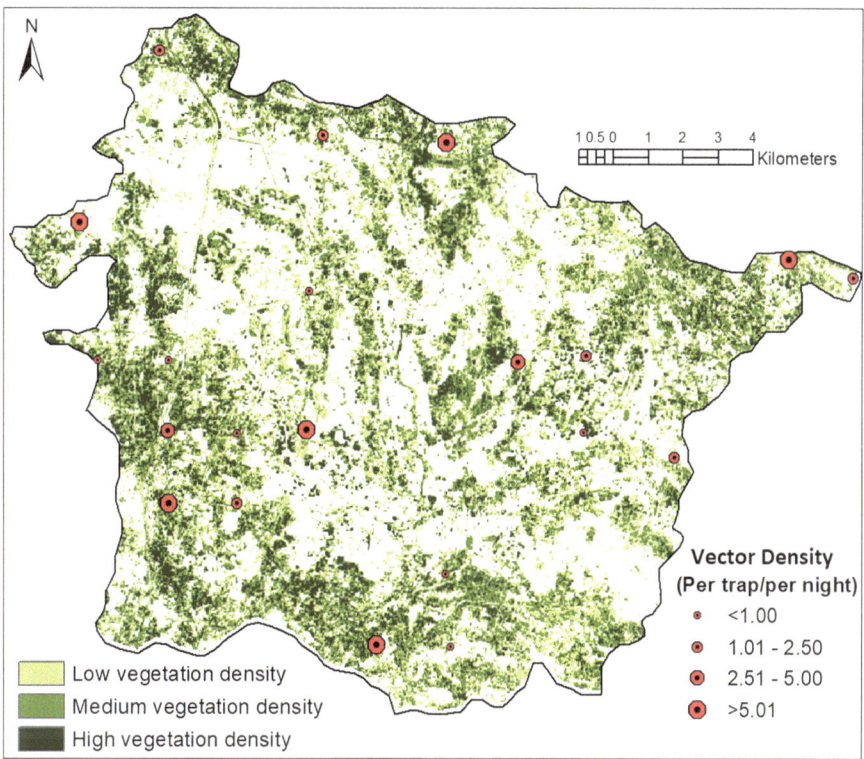

Fig. 4.6 Spatial association between Normalized Difference Vegetation Index and vector density. (Prepared in QGIS software)

Subsequently, a statistical relationship has been performed to understand the relationship between minimum and mean NDVI value with the *P. argentipes* density in Microsoft Excel (Fig. 4.7). Results showed a positive association with the minimum NDVI value and negative correlation with the mean NDVI value. The study implies that both the monthly variation of NDVI and sandfly density must be considered risk factors. These results can perhaps be attributed to the inclusion of areas with complex topographical and the climatic condition, diverse vegetation, and differing NDVI values in regional units providing sandfly density data.

4.7 Conclusion

GIS functionalities in public health analysis are still substantially underdeveloped in an Open Source GIS (OSG) software context in comparison with a desktop and commercial software application context. Majority of the OSG software use only choropleth maps and other cartograms like isopleth, graduated symbols to convey

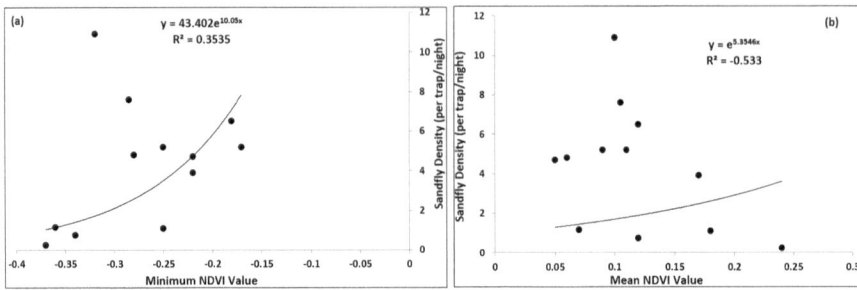

Fig. 4.7 Linear regression analysis between NDVI values and sandfly density. (**a**) Minimum NDVI value and sandfly density. (**b**) Mean NDVU value and sandfly density

health-related information. Beyond the traditional line, bar or pie charts and unconventional chart types have been used by other OSG software. Despite the practicality of the supplementary maps and charts, an interpretation tricky may ascend for end users without cartography background. Moreover, OSG softwares lag behind significantly in terms of providing users with robust methods of spatial and geostatistical analysis. Developing an effective OSG software requires collaborations of multiple disciplines. Integrating health-related data from multiple sources and enhancing the data analysis functions in Open Source GIS software to expand their recital are challenges that necessitate strenuous efforts. However, an appropriate balance still desires to be struck between Open Source software and secrecy matters. To lessen the resources essential for evolving effective OSG and providing useful information of health phenomena, espousing geomashup and open source models and firming local, regional, national, and global partnerships are essential.

Chapter 5
Sandfly Ecology of Kala-azar Transmission

Abstract This chapter focused on vector ecology on disease transmission. Approximately, 900 *phlebotomine* sandfly species have been recorded, and out of which 30% of the species are proven vector of kala-azar. Usually, sandflies are extended between 50°N and 40°S in the tropical and sub-tropical areas. Various ecological factors like rainfall, wind speed, relative humidity, soil moisture, pH, and organic carbon are known to affect the oviposition of gravid female sandflies along with the survival and growth of larvae. Sandfly density and behaviors of sandfly, risk mapping, and vector control strategy have been described. A case study has been described for identification of sandfly density using geographical factors.

Keywords Sandfly · Vector ecology · Kala-azar disease · Vector control strategy

5.1 Introduction

There are approximately 900 species of *phlebotomine* sandflies (Cheghabaleki et al. 2019), but less than 70 have been implicated in leishmaniasis transmission (Ready 2013). About 30% of species of sandflies are proven vector of at least 20 *Leishmania* species. The genus *Lutzomyia* occurs in the New World, and the species of the genera *Sergentomiya* and *Phlebotomus* are known to occur in the Old World (Alvar et al. 2012). The biology of each sandfly species is unique and multifaceted and has a through bearing in the epidemiology of leishmaniasis. Generally, the pathogen-vector-reservoir host relations are somewhat compound and have been exposed to vary spatially and temporally (Fig. 5.1). Several leishmania species blighting human are zoonotic (e.g., domestic and wild mammal reservoir hosts), while other species of the parasite are anthroponotic (e.g., human-to-human

Fig. 5.1 Vector ecology of
kala-azar transmission

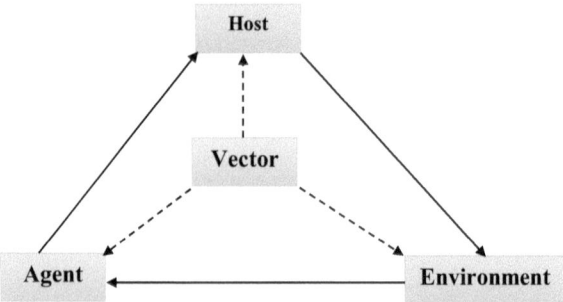

transmission in the presence of vector). In East Africa, transmission of *L. donovani* may take place through anthroponotic or zoonotic cycle (Elnaiem 2011). *Leishmania donovani* is usually considered to be an anthroponotic. In India, female *P. argentipes* are responsible for KA transmission, consisting of four life stages: eggs, larvae, pupae, and adults. Sandflies are vigorous at night and are stared as silent feeders (Poché et al. 2016). Parrot (1936) made the first description of *P. orientalis,* as a variety of *P. langeroni (P. langeroni orientalis)*. In 1955, Kirk and Lewis stated that *P. orientalis* is the main *L. donovani* vector in Sudan. The authors also suggested that *P. martini* and *P. lesley* may be the vectors of *L. donovani* in other areas of Sudan, where *P. orientalis* is absent. To complete the oviposition process, blood meals are essential for sandflies. In Europe and Latin America, the VL is zoonotic, and in Brazil it is caused by *L. infantum chagasi* which is transmitted to the humans and other animals by the bite of female *Lutzomyia longipalpis* (WHO 2010). Earlier report suggested that immature sandflies have been observed within and adjacent the cattle sheds, recommending cattle may assist as food source for larvae which feed on organic matter (Singh et al. 2008a). Every species of sandfly normally transmits just a single type of parasites, and every parasite prompts a specific kind of leishmaniasis form. Sandflies are taking blood mainly on cattle and humans within the rural and peri-urban areas. Since the leishmania parasites are transmitted through the chomps of tainted female phlebotomine sandflies, the epidemiology of leishmania species relies upon the traits of the parasite species, the nearby natural attributes of the transmission locales, cutting edge, and past exposure of the human populace to the parasite and human nature.

However, the spatial and temporal distribution of sandflies relies upon geographical and climatic factors. Normally, sandflies are found in the tropical and sub-tropical areas extended between 50°N and 40°S. As the sandflies are thermophilic in nature, they require high temperatures for their development and survival. Therefore, temperature is considered as essential factor for the development and survival of various stages, and thus it influences the geographical distribution of sandflies (Merino-Espinosa et al. 2016). In India, air temperatures at night in KA endemic region will exceed 20 °C in peak season (March–April and September–October), and in lean season (December–February), the minimum temperature are lowest, hence the sandfly population is lowest (Poché et al. 2016). Moreover, various ecological factors like rainfall, wind speed, relative humidity, soil moisture, pH, and organic carbon are known to influence the oviposition of gravid female sandflies along with the survival and development of larvae (Bhunia et al. 2010b; Chowdhury et al. 2016). Ambient temperature also affects the development rates of the immature stages, survival of pre-imaginal stages, and longevity of the adult phlebotomine sandflies (WHO 2010). In comparison to other mosquito-borne disease, sandflies do not lay egg in water; however, the sufficient moisture is needed for the survival of mosquito (Bhunia et al. 2011a). Moreover, the heavy rainfall can confine flight activity, bound resting site accessibility for adult sandflies, and destroy immature stages.

The information gaps in vector bionomics will be of instantaneous advantage to current control procedures including better approximations of human biting rate and natural infections rates of sandflies and how these differ geographically and temporarily. The spatiotemporal variation of environmental factors, such as temperature, humidity, and precipitation, affects the biology and ecology of vectors and immediate hosts and, consequently, the risk of disease transmission (Githeko et al. 2000). These climatic factors also affect the digestion, metabolic process, and developmental time of sandflies. Due to dearth of information on zoonotic transmission cycles and ecology of the vectors, little has been accomplished in menace and control of KA.

5.2 Sandfly Habitat and Risk Mapping

Elnaiem et al. (1998) suggested that in central Sudan, the presence of black cotton soil, rainfall, minimum and maximum NDVI, annual mean maximum temperature, minimum daily temperature, and soil types were found to be the best predictors of the presence and absence of *P. orientalis*. Elnaiem et al. (1999) recommended that high abundance of *P. orientalis* is connected with tree density, mound-building termites, and sources of sugar meals. Gebre-Michael et al. (2010) prepared a GIS risk map for *P. orientalis* based on the environmental variables and historical sandfly data in Ethiopia, Kenya, and Somalia. Calborn et al. (2008) reported that higher densities of sandfly were recorded in deciduous forest and at interface between forest and open grassland, whereas low sandfly densities were recorded from the evergreen forest and agricultural fields. Elnaiem (2011) reported that *P. orientalis* was encountered in *Acacia seyal–Balanites aegyptiaca* woodland area; however, this may be owing to the association of specific variation of the vector to these trees or as a result of the soil or other microclimatic features. The immediacy of vegetation types (e.g., *A. seyal*, *B. aegyptiaca*, *Combretum kordofanum*, *Hyphaene thebaica*, *and Ziziphus spina-christi*) and black cotton soil condensed the effect of confounding factors, other than vegetation parameters. Foley et al. (2012) customized SandflyMap using Microsoft Silverlight and ESRI's ArcGIS Server 10 software platform to delimit disease vector data and relevant remote sensing layers in an online GIS format. In the software modules, tools under development include:

- Mal-Area Calculator (MAC), which is raster-based and which takes into account spatial variation in the relative roles of different vectors and hosts and focuses vector hazard.
- Vector surveillance allows the display of temporal data.
- Vector Host Map to record pathogen and ecoparasite data from vertebrates.

The VECMAP software (https://business.esa.int/projects/vecmap) developed by AVIA-GIS, Belgium, provides a "One-Stop-Shop" by providing services like smartphone-based field sampling and data collection, statistical distribution modelling, and a centralized GIS and database. The software is supporting national public health agencies and regional mosquito controllers in European countries in envisaging vector-related health hazards and in plummeting feast of disease. Kesari et al. (2013) identified the habitat suitability of *P. argentipes* based on remotely sensed indices (e.g., land surface temperature and renormalized difference vegetation index) in India and state that nearly 74% accuracy of the model using these environmental variables. Cunze et al. (2019) provided climate suitability maps generated by means of an ecological niche modelling approach for 32 *Phlebotomus* vector species with proven or suspected vector competence for five leishmania pathogens occurring in Eurasia and Africa. They also used entropy modelling approach using Maxent software for predicting sandfly density based on the environmental variables.

5.3   Sandfly Ecology and Kala-azar Transmission

According to geographical distributional patterns of *L. donovani*, it might be classified into two groups:

 (i) Southern group – comprising *P. argentipes*, *P. martini*, *P. orientalis*, as well as *P. rodhaini*; they are mainly distributed in tropical areas and overlap with the distribution of *L. donovani*.
 (ii) Northern group – comprising *P. alexandri* and *P. longiductus*.

The relationship between *Phlebotomus* vectors and *Leishmania* pathogens are gathered from the literature, i.e., "confirmed" and "suspected" vector species (Fig. 5.2). In ISC, *P. argentipes* is the only vector of *L. donovani*.

Research on vector breeding sites are awkward and, therefore, scarce. Earlier study reported that human dwellings, tree holes, cattle shed, and mixed dwellings are positive for *P. argentipes* larvae and other sandfly species (Dhiman et al. 1983). Rahman et al. (1986) suggested that *P. argentipes* breeds in debris and the base of shrubs. However, on the basis of larval collections, it appears that *Phlebotomus* species prefers to breed in cattle sheds (e.g., associated with the alkaline soil) and human dwellings. Naturally, *Phlebotomine* sandfly are gonotropically concordant, captivating blood meal for each batch of egg maturation. Kesari et al. (2011) suggested that minimum normalized difference vegetation index, marshy land, orchard, and settlements showed high sandfly breeding in an endemic region. The moisture content of the soil is slightly higher and is suitable for vector habitation. In Nepal,

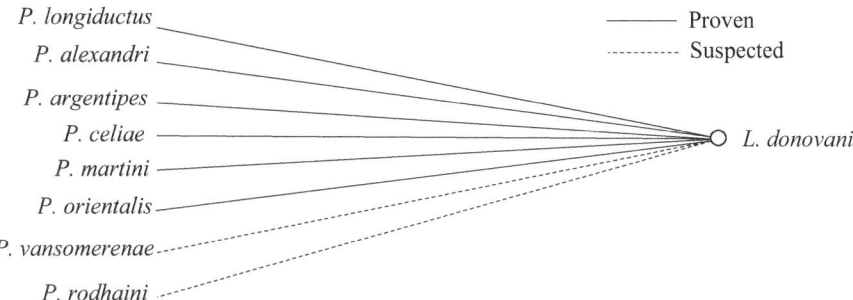

Fig. 5.2 Vector competence of *L. donovani* was confirmed (solid lines) and not yet confirmed but strongly suggested (dotted lines) according to WHO. (Source: modified after Cunze et al. 2019)

temperature from 20 to 24.9 °C had a highly significant effect on *P. argentipes* density (Das 2004). Most of the sandfly species probably feed on plant sugar and the female sandfly feed on mammalian blood (Kalra and Bang 1988).

Current evidence suggested that KA is strongly affected by ecological factors and also have a direct influence on disease transmission (Cheghabaleki et al. 2019). *Leishmania* parasites are transmitted by the bites of infected female *Phlebotomine* sandflies. The epidemiology of KA wings upon the characteristics of the parasite species. However, the local ecological characteristics of the transmission sites, current, and past exposure of the human populations to the parasite, and human behaviors. Air temperature, rainfall, pH, wind speed, humidity, soil moisture, and organic carbon are known to affect the oviposition of gravid female sandflies along with the survival and growth of larvae (Sivagnaname and Amalraj 1997). Earlier study reported that more than 88% of KA patients in India sleep outside for 5–8 months, coinciding with the peak density in *P. argentipes* abundance. The spatiotemporal distinction of environmental features influences the biology and ecology of vectors and intermediate hosts and, accordingly, the risk of disease transmission. Ecological investigations have established the role of spatiotemporal changeability on the sandfly's presence or activity patterns of several habitats. Habitat characteristics such as LULC were demonstrated to be important factors dictating activities on a regional or a macro-scale basis, with an emphasis on the distribution range of the species (Bhunia et al. 2012a; Waitz et al. 2019). Moreover, soil moisture, organic matter from agriculture, gardening, and waste may enhance conditions for both vectors and hosts (Waitz et al. 2019). With the change of climate and human-induced modifications of the landscape are incessantly taking place, along with the evolution of the human population, possibly altering the accessibility of habitats for host and disease vectors.

Numerous investigations have addressed the effects of specific breeding locations, resting and foraging sites, and various ecological features on the sandfly activity (Bhunia 2014; Kesari et al. 2011; Sudhakar et al. 2006). The density of sandflies in the ecotone would be reliable with the obtainable blood source in spite of the perpetual source of blood in the zones of altered landscape. While the spreading distances of *Phlebotomus* spp. are considered to be restricted to a few hundred meters, some species parade higher mobility, up to 2 km (Orshan et al. 2016).

Variances in movement patterns and remoteness have been found between the sexes, where males have a tendency to amass in the host's habitats (e.g., forest or human environments), caring a discrete territory near the blood source.

5.4   Sandfly Density and Potential Reservoir and Hosts

Sandflies occur in all seasons with small as well as large abundance peaks. The main peaks of abundance occurred in the post-monsoon season which is separated by the pre-monsoon season. However, the seasonal fluctuations in sandfly density showed peaks in November and during the period between April and May. The appearance of adult *P. argentipes* from soil samples was detected between April and October (Singh et al. 2008a, b). Poché et al. (2011) reported the highest proportions of gravid female sandflies during May. In Bangladesh, the highest sandfly density is recorded in March and the lowest sandfly density is recorded in December–February (Chowdhury et al. 2011). The seasonality of *P. argentipes* showed two peaks (i) between April and May and (ii) between September and October. The sandfly distribution is affected by climatic and environmental factors and must concur in space and time with the dispersal of infections in humans (Bhunia 2014). Most of the sandflies are caught in cattle, and the humans are also found to be the preferred host. During the post dichloro-diphenyl-trichloro-ethane (DDT) era, species showed increased zoophily which was accredited to alteration toward cattle by prickliness to DDT. Subsequently, sampling either from cowsheds or human habitation gave variable results, indicating that host preference of *P. argentipes* varies widely in different biotopes. In Nepal, more *P. argentipes* are collected from cattle sheds than the human dwelling. In West Bengal, blood index showed much higher in cattle in respect to human (Addy et al. 1983) and it is equal to the *bovid blood index* (e.g., proportion of bloodmeals of a sandfly population obtained from cattle). Chowdhury et al. (2016) reported that *Phlebotomine* sandflies are mainly zoophilic and exophilic and are collected from the living rooms of houses in both the rural and urban areas, but fed on humans in the absence of bovines. Garlapati et al. (2012) stated that sandflies mostly fed on humans, followed by cattle, buffalo, and goats. It appeared that the host preference of *P. argentipes* varied widely in different biotopes, usually small or well-defined area that is uniform in environmental conditions but was mainly zoophilic, preferring to feed on human as the second choice; however, it is also a "chance feeder" according to biotopes (e.g., distribution of plant and animal life).

5.5   Sandfly Behavior and Risk of Infection

Underdeveloped houses (e.g., mud and thatched houses), peridomestic environment, garbage distribution surrounding the houses, open sewerage, poverty, malnutrition, and change in climatic condition such as temperature, rainfall, and humidity may increase sandfly breeding and resting sites at household level. Moreover, sandflies are fascinated to crowded housing owing to a greater number of hosts (WHO 2010). Moreover, transmission can be zoonotic or anthroponotic depending upon the reservoir. Domestic animals and cattle population in endemic regions hold epidemiological importance as they take part in the transmission of the parasite by serving as reservoirs. In India, humans with VL or PKDL serve as the only reservoir in anthroponotic transmission. Additionally, parasite-vector contact is rare for great majority of the sandfly species (Ready 2000). In many areas, the intrusion of humans into sylvatic cycles increases the risk of infection. Human being infers this cycle in search of agricultural land, human settlement, timber production, road construction, and other economic benefits in forest and other enzootic areas (Alemayehu and Alemayehu 2017). In relation to human settlement, the houses nearer to agricultural villages are at higher risk for leishmaniasis transmission, as the houses are constructed with mud walls and earthen floor along with mixed dwelling. Earlier study also reported that sleeping outside in warmer months is associated with the increase of risk of infection. In disease-prone areas, houses nearer to natural habitats of the vector and reservoir have increased risk of infection. Moreover, migration of population from endemic rural environment into urban and peri-urban areas is a major reason for the establishment of disease transmission (Quinnell and Courtenay 2009).

5.6 Vector Control Strategy

Detailed knowledge of seasonal variation of sandfly densities, biting time, and biting places as well as host preferences can improve our understanding of prospects and limitations of certain interventions and direct us to tailor effective control operations. The difficulty of delimiting breeding sites is a significant restraint on vector control measures. Till date, there is a dearth of detailed information on the life cycle of *P. argentipes*. Indoor residual spray (IRS) is very effective in reducing sandflies under strict supervision, but very limited information is available on the impact of IRS on VL transmission. Moreover, the outcome of IRS did not achieve optimum results when applied by national control programs (Chowdhury et al. 2010). The use of long-lasting insecticide-treated nets (LLIN) is another method of vector control in National Inter-ministerial Steering Committee (ISC). However, on the basis of bioassay criteria, all LLINs tested met the WHO Pesticide Evaluation Scheme standard for LLIN effectiveness (Picado et al. 2011), and recommending that untreated bednets run some degree of special guard against VL. However, the current available information regarding insecticide susceptibility for controlling VL vectors which would guide vector control services to adjust the insecticides in use lacks updating. Alternatively, plastering of walls and floors using mud and lime plaster are associated with the decrease of *P. argentipes* to control. Recently, researchers and scientist have made efforts in using satellite data in predicting vector density. Several vector control models and early warning system that employ EO data have been used as tools for serving decision-makers to recover health system responses, yield pre-emptive actions so as to restrain the expansion of sandfly presence or absence, and discourse the pertinent primacies of the sustainable development.

5.7   Case Study: Identification of Sandfly Density Based on Geographical Factors: In Situ Observation and Geospatial Technology

5.7.1   Introduction

The distribution of the sandfly is determined by various key environmental and socioeconomic variables; thus, the spatiotemporal pattern of VL is considered in relation to those covariates (Ding et al. 2019). Outside the geographical limits of natural foci, the abundance of sandfly is increasing and leads to the modifications in the classic epidemiological patterns. Such modifications are associated with the demographic expansion, environmental modification, and climate change. Climate change persuaded extension of vector-competent species would at the same time lead to a growth for VL-risk area if disease diffusing *L. donovani* parasites are capable of live and start in expanding vector populations. Although there have been numerous studies conducted by several researchers on phlebotomine species distribution (Bhunia et al. 2011a, b; Abdullah et al. 2017; Mandal et al. 2018), an extensive modelling approach including several sandfly species is still missing. The existence of the vector species is observed as a predisposing aspect for occurrences of leishmaniasis in tropical areas. There are several factors, mainly the distribution of the associated parasites and its growing environments that should be considered for future modelling and risk valuation (Koch et al. 2017). The ecological niche model (ENM) has been used widely as a tool to describe environmental conditioning factors and to identify the patterns related to environmental suitability for species occurrence (Peterson et al. 2011; da Costa et al. 2018). Presently several techniques have been adopted for modelling niches and species distributions. Therefore, in this study, a collective predicting approach is considered to estimate the habitat suitability of *P. argentipes* distribution based on ground-based observation and geospatial technology.

5.7.2   Environmental Indicators

Vegetation plays an important role in sandfly habitat and survival. It provides the necessary sugar meal and maintaining the moisture profile for sandflies. Moreover, vegetation could reduce evaporation, decrease wind speed, and protect certain areas from direct sunlight which is suitable for survival of dipterans (Bhatt et al. 2013). Bhunia et al. (2012a, b, c) have also reported that there is a link between VL and kala-azar. In this study, land cover was adopted as a key explanatory variable in the distribution of VL cases. Sandflies are thermophilic in nature. Earlier study reported that there is a link between terrain and KA (Bhunia et al. 2014). Although the relationship has not been understood, we assumed that topography may restrict the vector to certain geographical areas. In this study, an elevation dataset generated by the SRTM was used as good measure for topography. In this research, environmental indicators such as, elevation, surface wetness, land surface temperature, vegetation cover, land use/land cover characteristics are considered. Moreover, a spatial correlation is made between each environmental variable and sandfly density. The sandfly population with higher density is marked as high-risk for sandfly abundance and vice versa.

5.7.3 Socioeconomic Indicators

There is a strong and complex association between KA and socioeconomic covariates. Boelaert et al. (2009) demonstrated that impoverished populations are most prone to VL, as poor housing conditions and detrimental habitats increase sandfly breeding and resting sites. Furthermore, poverty is associated with the poor nutrition which conciliations the insusceptibility of poor populations, education level, peridomestic condition, spraying around houses, types of house wall, presence of greenery inside the houses, irregular spraying around houses and increase the menace that VL infection will growth to the clinically established disease (Ranjan et al. 2005). Bern et al. investigated the geographical pattern of Bangladesh and related risk features like closeness to KA patients, number of cattle per unit area, age and sex, napping situations, eating habits, income level, and use of bed nets. In this study, socioeconomic indicators, such as cattle density, mixed dwelling, underdeveloped houses, and agricultural labor, are considered for the sandfly abundance. A spatial overlay analysis is performed to understand the relation between socioeconomic indicators and environmental variables.

5.7.4 Climatological Indicators

Climatic factors (temperature, precipitation, and humidity) have a strong effect on the ecology of vectors and reservoir hosts by influencing their survival and breeding and resting of vector. Temperature is a crucial issue for the expansion and subsistence of various life stages, and thus, its stimuluses spatial dispersal of sandflies. Moreover, sufficient moisture is still important for egg survival (Kasap and Alten 2006). Moreover, heavy rainfall can restrict flight activity, limit resting site availability for adult sandflies, and kill immature stages (Bhunia et al. 2010a, b). Due to change in climate, several sandfly species are expected to expand their dispersal ability. In situ temperature and relative humidity are collected from the sandfly collection sites, and these data are integrated into GIS platform. Radial basis function (RBF) interpolation technique is considered for delineating climatic condition for the entire study area. The sandfly density data overlay on the interpolated map and the threshold value of temperature and relative humidity is identified for the sandfly density.

5.7.5 Statistical Analysis

All the obtained data are analyzed by using Microsoft Excel version 7.0. To identify the relationship between the sandfly density and environmental variables, a univariate analysis (Pearson correlation co-efficient) is performed. Moreover, a multivariate analysis is performed to predict about the presence of sandfly density.

5.7.6 Formulation of Risk Model

The weighted score analysis is considered to generate the sandfly density. The weighted score (W_i) is calculated for every significant ($P < 0.05$) environmental variables (E_v) derived from the univariate logistics regression analysis as $W_i (1 - p_i)$. Considering the spatial association on the sandfly abundance, the weights of different themes are allotted on a scale of 1–5. These scores are used to identify the density values (R_v). The overall methodology of the study is represented in Fig. 5.3. Each variable is analyzed in GIS platform and weightage is assigned for each subcategory. Reclassification analysis has been done for category of environmental, socioeconomic, and climatological indicators. Subsequently, all these indicators are integrated into GIS platform and final output is prepared. The density value for sandfly species is individualistically demonstrated as follows:

$$R_v = 1 / \log\left(w_1 E_{v_1} + w_2 E_{v_2} + \ldots\ldots + w_n E_{v_n} \right)$$

5.8 Results and Discussion

The intrinsic factors are the biochemical and physiographical properties of the sandfly that determine its reaction to the external conditions of its habitat. The wider use of remote sensing data sets for risk assessment is needed to better understand the potential distribution of the vector in an endemic kala-azar focus. The value of NDVI varies from 0.17 to 0.43 (mean ± standard deviation: 0.18 ± 0.02). However, the higher value of NDVI indicated that the vegetation density of this region is high and vice versa. The higher NDVI value is recorded in the southwest and west part of the block, while the lowest value is evident in the southern part of the district. Results showed that areas covered by less vegetation are at high risk for sandfly abundance. A negative correlation is found with mean NDVI ($r = -0.56, p < 0.002$) and maximum NDVI ($r = -0.45, p < 0.020$) and sandfly density. The LST ranged from 23.00 to 36.00 °C (mean ± standard deviation: 26.50 °C ± 4.76). The strong and positive correlation also existed between sandfly density and the minimum LST ($r = 0.65, p < 0.026$), followed by mean LST values ($r = 0.64, p < 0.016$). The study area is classified into river, settlement, surface water body, moist fallow, crop land,

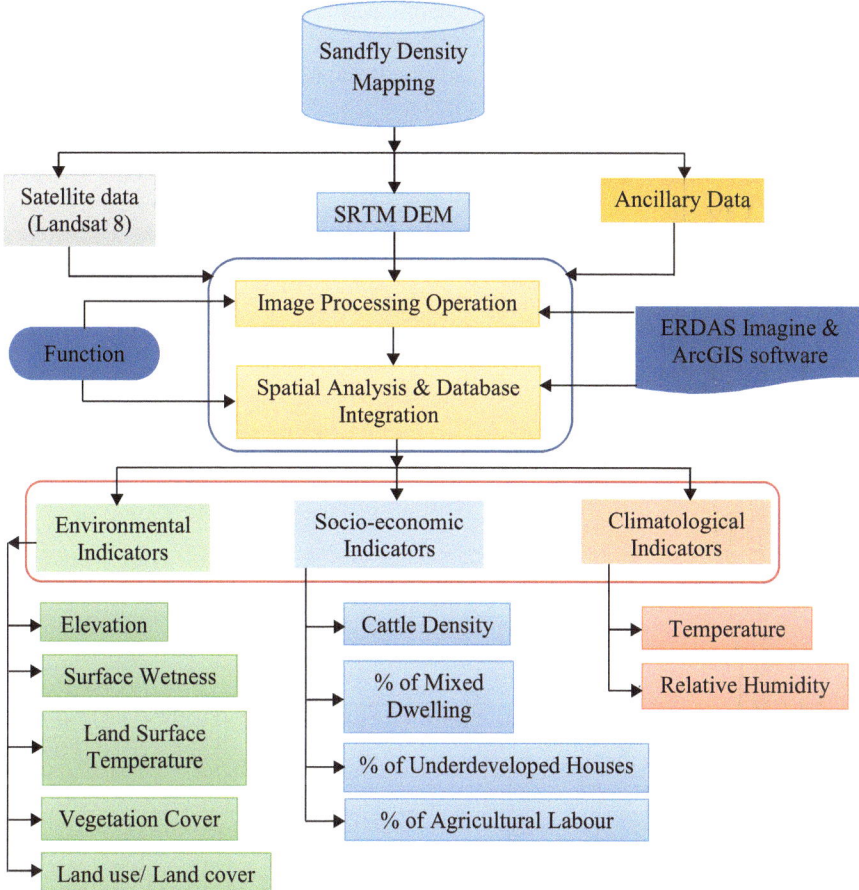

Fig. 5.3 Flowchart map of sandfly density mapping

vegetation, and plantation with settlement. A wetness index (WI) map was prepared to investigate the surface dampness of study area. Pearson correlation tests showed that there is a significant negative relationship with maximum and mean WI values with *P. argentipes* abundance ($r = 0.48$; *P < 0.003*). Overlaying the sandfly density on WI map showed that areas with WI values of −7.68 to 6.29 usually corresponded with areas with high numbers of sandfly density.

Underdeveloped house is a significant parameter of the model. Most of the sandfly are recorded from the mud and thatched houses in the study site. This seems to be reasonable as the people spend most of their time at home and increased the chances of sandfly bites and disease transmission. Agricultural labor types are also considered as significant variables in the study area. As these workers spend more time at home in peridomestic and unhygienic areas where vectors are present, and more likely to contact the disease. Moreover, mixed dwelling and cattle shed nearer to the houses also play an important role for the sandfly abundance and increase the risk of transmission of disease. The maximum number of vectors is recorded with the temperature value ranging between 26.00 and 30.00 °C. The simple linear regression analysis showed a strong and significant relationship ($r = 0.58$, $p < 0.05$). Moreover, if the temperature is too hot and dry, vector survival rate will lessen and the disease may vanish from some areas. However, the largest number of sandfly *is* recorded with the relative humidity of >70.00%. Pearson correlation coefficient test showed strong positive correlation ($r = 0.67$, $P < 0.05$). Climate change influences the spreading range of vectors and pathogen. The higher seasonal temperatures would tend to protracted movement periods and shorter diapause phases that may also increase number of sandfly generation per year. For example, climatic parameters increase it hastens maturation of leishmania parasite, and also escalate the risk of infection.

Probable sandfly density (per trap/per night) map of Jandaha block of Vaishali district is prepared (Fig. 5.4). The highest sandfly density (~10.34 per trap/per night) is represented in "red" color and the lowest sandfly density is represented in "pink" color. The highest density is mainly observed in the eastern and south east of the Paroo block. The lowest density (~6.0 per trap/per night) is observed in the southwest and small pockets of north-west corner of the block. The central part of the study site illustrated medium to high density of sandfly, represented in "yellow" color. The dispersal of *phlebotomine* sandfly is increasing in environment outside

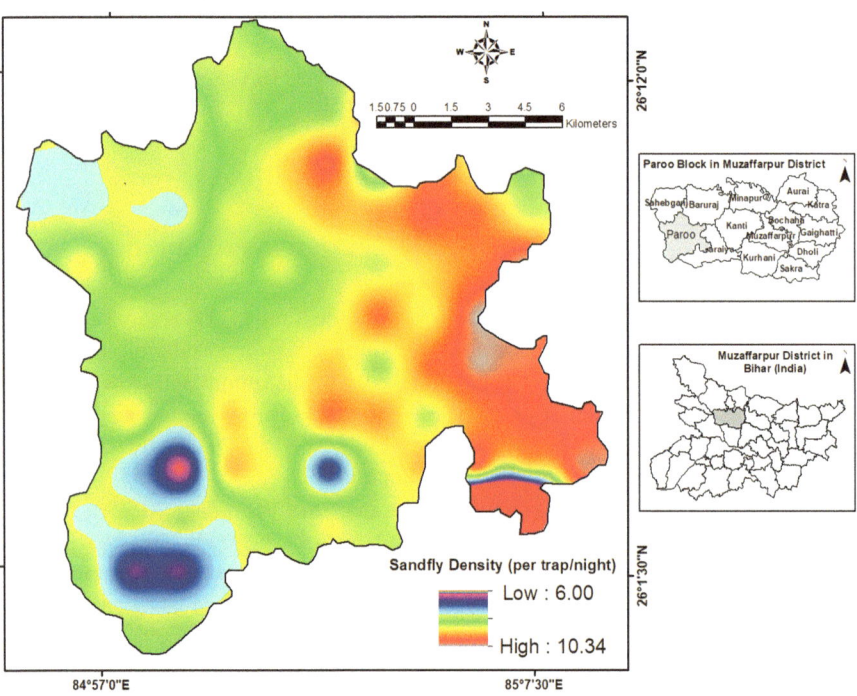

Fig. 5.4 Sandfly density map of Paroo block in Muzaffarpur district in Bihar (India)

the geographical limits of natural foci, which, hence, leads to modifications in the classic epidemiological patterns of leishmaniasis. Such variations are associated with changes in the predisposing factors for the exposure of human to transmission, demographic extension, and the process of suburbanization on the boundaries of natural foci, in addition to the incidence of periurban and forest remnant areas (da Costa et al. 2018).

5.9 Conclusion

However, this study showed that local geographical factors play an important role in the transmission of disease. Moreover, remotely sensed environmental variables and ground-based observation of climatic and socioeconomic characteristics allowed the prediction of sandfly density at micro level. Undoubtedly, there are numerous interconnected ecological factors that require to be considered for the prediction of sandfly. The uses of LULC characteristics facilitate the biological interpretation and by investigating whether particular land-cover types are associated with sandfly abundance along with the surrounding peridomestic environment. Moreover, the vegetation index may be used as a secondary variable for prediction. Results also showed mean and minimum LST has a significant effect on sandfly abundance. These results imply that temperature and relative humidity might be the determinant factor in forecasting sandfly abundance, such as mating and oviposition in the VL endemic area.

Chapter 6
Disease Ecology and Transmission

Abstract This chapter describes about the ecological factors in kala-azar transmission mapping and modelling. Risks of emergence or re-emergence of leishmaniasis are associated with several aspects. The coinfection of disease with post-kala-azar dermal leishmaniasis (PKDL), tuberculosis, and HIV/AID has also been analyzed. Overlap of HIV/VL coinfection is a serious worldwide concern. Moreover, changes in landscape variables, increasing agricultural land, forested areas to domestic and rural settlements, etc. have been found to be associated with the VL risk. A correlative analysis between environmental factors and disease risk analysis has been described. Finally, a case study on risk zone mapping and modelling of kala-azar using environmental factors via geospatial tool and field-based observation has been described. Weighted index overlay analysis is used to identify the kala-azar risk areas.

Keywords Kala-azar · HIV/VL coinfection · Post-kala-azar dermal leishmaniasis · Disease risk mapping

6.1 Introduction

Visceral leishmaniasis (VL) or kala-azar (KA) is closely linked with the natural environments and has been regularly cited as a dispute on health risk associated with the change and instability of the climate. VL-infected human cases can be categorized as asymptomatic, symptomatic, and post-kala-azar dermal leishmaniasis (PKDL)-infected representing various stages of disease diagnosis. Five countries

G. S. Bhunia, P. K. Shit, *Spatial Mapping and Modelling for Kala-azar Disease*,
SpringerBriefs in Medical Earth Sciences, https://doi.org/10.1007/978-3-030-41227-2_6

(Bangladesh, India, Nepal, Sudan, and Brazil) account for 90% of the global VL burden. Approximately, 147 million people are at risk in South Asia region. In China, there are three defined VL types: anthroponotic VL (e.g., caused by *L. donovani*), zoonotic mountain VL, and zoonotic desert VL (e.g., caused by *L. infantum*). Among East African countries, Ethiopia, Sudan, and Somalia have maximum number of VL case report. In Europe, countries like Italy, Greece, Turkey, and Albania have highest prevalence of VL cases. In America, Brazil has the highest number of reported cases. Since the mid-1990s, the worldwide geographical distribution of leishmaniasis has expanded. A number of species in the genus *Phlebotomus* are responsible for transmission of VL in the Old World and *Lutzomyia* in the New World. Several species of wild, domestic, and synanthropic mammals have been recorded as hosts and/or reservoirs of *Leishmania* spp. in various parts of the world. It is reported that human beings are directly involved as a principal reservoir host for VL. Environmental factors acquired from various sources, including geographic information and remotely sensed data, have been used to predict and elucidate the distribution of the disease caused by vectors. The demographic aspects like new agro-industrial activity, large-scale migration, unplanned urbanization, and modification of the physiographical characteristics also influence the disease distribution. Since the last decade, ecological niche models (ENMs) based on point occurrence data, environmental layers, and machine learning algorithms, typically all overlaid on a GIS, provide a useful framework for understanding the geography of KA. This approach has proven to be highly useful for disease prevention and has been used to forecast epidemics, which is imperative for the preparedness of health systems to cope with such outbreaks. Subsequently, the natural environment is constantly disturbed by humans through agricultural development, deforestation, ever-increasing population, etc. Minimizing "ecological risk" of VL spread could be an important alternative to keep its occurrence at an acceptable level.

The reported VL cases have been increased globally in the last two to three decades. HIV/VL coinfection, PKDL, and drug resistance over time make it high-risk globally (Ready 2010). Since the last decades, the launch of the Regional Strategic Framework for VL elimination in 2005 has shown a substantial decrease of VL (more than 75%) in the Indian subcontinent. There is a more recent initiative by WHO to define a road map for deterrence and control of VL and annihilation of VL by 2030 as a step toward attaining the Sustainable Development Goals. Because of the new and complex epidemiological scenarios, VL is considered a reemerging disease (WHO 2016). Moreover, the disease shows widespread geographical distribution, and new cases are recorded from some areas which are nonendemic previously. Negative ecological interaction and dispersal behavior may prevent the species from occupying the entirety of its fundamental niche. ENMs incorporate both the ecological variables and epidemiological and entomological characteristics, and predict the species occurrences in an area between the potential and actual distribution. Since the last decade, several numbers of published literature have focused on the paraphernalia of biodiversity on the risk of KA occurrence. Hence, understanding the structure and functioning of the ecological processes involved in the dynamics of the interaction between parasite, hosts, and environments. The vulgarization of the living communities and the crumbling and forfeiture of habitats ensuing from human occupation alter the vector/host relations which may lead to the emergence and re-emergence of KA. Various modelling techniques are now available to integrate environmental layers with disease cases, which allow environmental factors to be isolated and potential vector distribution to be mapped.

6.2 Post-Kala-azar Dermal Leishmaniasis

Post-kala-azar dermal leishmaniasis (PKDL) is an identified problem of VL caused by *L. donovani*, but rarely caused by *L. infantum* and *L. chagasi*, and is possibly the most intriguing clinically and scientifically, as it usually progresses as a sequela after apparent effective cure from VL (El Hassan et al. 2013). The clinical representation of VL and PKDL varies considerably. In VL, patients ache from prolonged fever, hepatosplenomegaly, weight loss, and anemia, whereas the appearances of PKDL are limited to macules, papules, or nodules involving the skin, particularly in the sun-exposed areas. PKDL lesions are not similar to cutaneous leishmaniasis and are volcanic in shape, have fixed peri-oral spreading, and are associated with hyperpigmented macules (El-Hassan et al. 1990). PKDL is confined to South Asia (India, Nepal, and Bangladesh) and East Africa, mainly Sudan (Desjeux and Ramesh 2011). In ISC, polymorphic lesions (e.g., macules, papulonodules) are prevalent, whereas Sudanese variant has popular or nodular lesions. Though PKDL is a stigmatizing disease, its mortality rate is less. However, understanding the clinic-epidemiological aspects of PKDL would help define strategies for controlling VL, by providing further insights into *L. donovani* transmission dynamics. In South Asia, transmission of VL is anthroponotic, while in Sudan, it is zoonotic and anthroponotic. In India, PKDL patients are proposed reservoirs for VL transmission. Accordingly, eradication of PKDL should be an essential component of the current VL elimination program in South Asia that aims to bring down the annual incidence of VL.

6.3 Kala-azar and Tuberculosis Coinfection

Nowadays, Kala-azar/VL and pulmonary tuberculosis (TB) coinfections are an increasing public health issue, especially in developing countries (Dye and Williams 2010). VL can produce latent TB. More than 80% of TB cases and deaths are reported from developing countries, and the TB conditions are intensified by high occurrence of HIV, concurrent with other parasitic disease. Since the last 70 years, several studies have reported the coinfection of TB and VL and few studies showed that the immune response was modified in the coinfection situation (Zaman et al. 2006). VL/TB coinfection has important clinical implication and shares numerous features, and several infections persist asymptomatic. Symptoms usually develop after several months due to very long incubation period and are associated with the immune suppression, occurred at the last stage. In eastern region of India, TB/VL coinfection increases dramatically, although their etiology and transmission mechanism are different, but they have shared several features. Moreover, the treatment of VL depends on the development of effective immune response which activates macrophages to produce nitric oxide for killing intracellular amastigotes.

6.4 HIV and Kala-azar Coinfection

Leishmania/human immunodeficiency virus (HIV) coinfection has global distribution and has been documented in 35 countries. Before 1990, HIV/VL coinfection was noticed in the Mediterranean basin. This phenomenon has happened predominantly in coastal regions of Spain, France, and Italy (Alvar et al. 2008), where official incidence of VL cases almost doubled during the period between 1987 and 2004. In India, HIV/VL coinfection is low (<1%). In India, HIV/VL coinfection is an underdiagnosed and under-recognized emerging public health problem in the Indian framework, necessitating urgent consideration. The key aspect of HIV/VL coinfection is that parasitic infection may persuade the stimulation of latent virus. Clinical researches have recommended that leishmaniasis endorses a growth in serum HIV type-I lead and a more rapid development to AIDS that lessen life expectancy in HIV-infected patients. One of the major contests of HIV/VL coinfection is emergent an effective drug therapy that not only tenacities the first episode of VL, but also avert relapse. HIV/VL coinfection hinders therapeutic response and is the cause of frequent relapse, especially in patients with CD4<200 cells/μl. Very few clinical trials have been conducted on the efficacy of some therapies for HIV/VL coinfection; the majority has been conducted in Europe and East Africa. VL/HIV coinfection leads to death. HIV/VL coinfected patients are instigated by recrudescence of a hidden infection which turns into clinically deceptive as the immunosuppression progresses and parasite overwhelms the eventuality measurements of host's immune system (Morales et al. 2002). PKDL is common in HIV-negative patients, appearing shortly during and after the treatment. Very few cases of PKDL in HIV-positive patients have been reported. In Ethiopia, PKDL was more common in HIV-positive patients. An association between PKDL and immune reconstitution disease (IRD) soon after start of highly active antiretroviral therapy (HAART).

6.5 Correlative Analysis of Microenvironmental Factors and Kala-azar Transmission

The development and incidence of VL are principally reliant upon ecological aspects and natural environments. Moreover, the economic, social, and cultural conditions also play an important role in disease transmission (Bhunia 2014). Earlier report also suggested that forest fragmentation, climatic variables, vegetation vigor, land surface temperature derived through remotely sensed data are the important predictors for VL (Bhunia et al. 2011b). Moreover, changes in landscape variables, increasing agricultural land, forestial areas to domestic and rural settlements, etc. have been found to be associated with the VL risk. In European countries, seroprevalence data shows high endemicity. Subsequently, migration, climate change, cattle population, and coinfection with HIV and tuberculosis are the main driving factors that increased incidence and prevalence of VL (Oryan and Akbari 2016). Moreover, the transmissibility ability of the reservoir host species and approachability of the *L. donovani* to the vector are significant for preservation of leishmania transmission in an area. This may be attributed to a combination of numerous aspects, which comprise many ecological factors (e.g., bio-climate, soil, vegetation, etc.), socioeconomic characteristics, land use/land cover characteristics and low efficiency of VL control program that generate new habitats in microclimate, and variations in seasonal climate. To address changes in their preferred bio-climate niches driven by climate change, phlebotomine species, as well as zoonic reservoir, can spread to a new suitable habitat. However, these movements are limited by their own intrinsic biological resilience and the potential of these vectors and reservoirs.

Visceral leishmaniasis (VL) generally tends to affect the lowliest people and downgraded societies. People who live in humid houses, locality of accrued rubbish, mixed dwelling, and underdeveloped houses (e.g., mud or thatched house) have been found at risk for VL (Kesari et al. 2010). The domestic animals and cattle population in proximity of the houses has been emphasized the risk of VL. Moreover, infected individuals may play an important role in sustaining VL and result in endemicity in poor communities. Picado et al. (2014) reported that worse socioeconomic status is associated with an increased risk of seroconversion, which is strongly associated with the leishmaniasis. Finally, study of the risk factors for this disease is multifaceted due to unstable transmission and high number of latent infections.

6.6 Kala-azar Control Efforts

Control of VL/KA is not easy in consequence of (i) ensues in various ecological situations, (ii) restraint of diagnostic method which are hostile, affluent and necessitates multifaceted infrastructure, (iii) restraints of the therapeutic interferences with an improper frequency and strength of adverse effects. The control strategy for KA/VL is depend upon the hostile documentation of cases, operative management and vector control measures for lessen of, not only morbidity and mortality, but also disease spread. The usage of simple, consistent, and low-cost tests for field-level serological diagnosis, new orally managed drugs, and long-lasting insecticide bed nets is predicted to significantly diminutize the number of cases, lessen transmission, and prevent epidemics. During the period between 1991 and 1992, an enhanced effort by the government of India coalescing greater obtainability and easy convenience of drugs, and spraying of houses with DDT. With the introduction of the disease control program in 2005, there had been a sharp decrease in the number of cases . Environmental vector management was implemented as part of the elimination program through insecticide spraying, utilization of bed net, etc., but its effectiveness showed mixed outcomes. Thornton et al. (2010) emphasized that a "one-size-fits-all" strategy may not be an appropriate approach because VL occurrence is a multifaceted problem. Joshi et al. (2008) stated that the complete elimination may be difficult to achieve in South Asia due to a number of reasons, including shortcomings of disease surveillance system and resource constraints.

6.7 Case Study: Kala-azar Epidemic for Vulnerability Zones – Multicriteria Approach

Visceral leishmaniasis (VL) or kala-azar (KA), also known as black fever or Dumdum fever, is caused by *L. donovani* and *L. infantum* and has a broad distribution throughout many temperate, subtropical, and tropical areas of the world. Phlebotomine sandflies are the prime vector for VL, which primarily inhabit in hot and wet tropical regions. However, sometimes they also inhabit in dry and wet tropical regions. The cause of epidemic and the transmission of disease are influenced by the geographical and environmental variables. Geographically, the universal variables represent the geographic area where the vector has its niches, and climate also plays an important role for the growth of vector and the abundance of sandfly that are relatively propitious to the transmission of disease. Moreover, the anthropomorphic factors and climate change in a specific region are associated with the VL occurrence. In terms of mortality and morbidity, the fatal parasitic disease was ranked ninth in a global analysis of infectious disease by the World Health Organization. The connotation of the vector with natural reservoir (e.g., canine, chicken, bovine, equine, ovine, swine, feline, etc.) turns into a favorable aspect toward keeping an endemic status for VL. The epidemiology of KA relies upon the coexistence and interaction of the vector, host, and parasite. The local environmental factors of the transmission sites, past exposure of human population to the parasite, and present human behavior determine the infection status in humans (Bhunia et al. 2010a, b). The disease tends to affect the poorest of the poor people, particularly those people who are close to water resources, humid environment, mud/thatched room with cracked walls, and low socioeconomic status (Oryan and Akbari 2016). Several numbers of experimental and theoretical approaches are provided by researchers and scientists to detect the spread of VL infection and control strategy. From the last few decades of the ninth century, mathematical and statistical model of infection transmission has provided a greater understanding in the solution to curb the spread of leishmaniasis transmission and its incidence in populations. Most of the models attempted to investigate zoonotic VL disease transmission through population (Burattini et al. 1998).

High-resolution satellite imagery provides integrated landscape characteristics and spatial distribution of earth surface features, with spatial relationship among the surrounding environment. The combined use of GIS and statistical techniques, incorporating a combination of geo-environmental parameters, allows the identification of risk factors and geo-localization of risk areas. GIS is knottily interrelated with several interfaces of human, data, server, and tools and comprises a set of plans and tools proficient in assimilating, storing, editing, investigating, and demonstrating spatially referenced information from numerous sources. Hence, an attempt has been made to develop a geo-environmental risk model in relation to kala-azar disease transmission based on primary and secondary data with the aid of remote sensing and GIS technologies.

6.7.1 Sampling of Adult Sandfly

The studies of sandflies are essential to determine the fauna, their distribution, and population dynamics involved in leishmania transmission and ultimately for building up cost-effective control methods for ultimate goal of disease control. Use of Communicable Disease Centre (CDC) light traps sandflies are collected from the 20 sample sites randomly in the post-monsoon season. Sandflies are attracted to the CDC light trap and sucked into a cage by a small fan. CDC light traps are placed 50–70 cm above the ground and are run (between 18.00 hours and 06.00 hours) once a month, yearly. The locations of traps are selected randomly in high-endemic villages in the block. For each village, a minimum of ten houses are selected to collect the sandflies data. All the sandfly species are swelled on Canada balsam microslides (Remaudière 1992). For species recognition, Lewis (1978) is followed. The sandfly density was measured through the number of sandflies collected on a given night and divided by the number of traps set out on that night.

6.7.2 Calculation of Incidence Rate

It should be noted that clinically diagnosed and laboratory-confirmed human infection cases reported during 2013–2015 are used in the modelling process and suspected cases of VL are not used in this study because of their own uncertainty. VL incidence rate is calculated for each village based on the mid-year population. The geoposition information on these cases is at least accurate at the township level, and most can be detailed at the village level. By combining Google Earth with the geopositioning information of the cases, VL occurrences were manually geopositioned to the point level, with the coordinates checked to ensure that they are plausible.

6.7.3 Ground-Based Observation and Assessment of Climatic Factors

Temperature and relative humidity (RH) are collected from the field during the survey. The climatic data are collected from the 25 sample sites. Based on its geographic location, point layer is prepared in GIS platform and the climatic variables are integrated into spatial data. Finally, RBF interpolation technique is used to understand the spatial distribution of climatic characteristics in the entire study area.

6.7.4 Identification of Topographic Variables

To identify the topographic variables, the entire study area is divided into 1 × 1 sq. km grid area. The minimum and maximum elevation for each grid is calculated from the Google elevation profile. Based on this information, absolute (e.g., maximum elevation recorded within the grid area) and relative relief (e.g., the difference between maximum and minimum elevation within the grid area) is calculated and interpolated for the entire study area. Finally, the past record of disease incidence is overlaid on the topographic variables, and threshold value for the topographic parameters has been identified for disease risk.

6.7.5 Environmental Variables

Vegetation, surface wetness, land surface temperature, and land use/land cover characteristics are important candidate predictors for KA transmission. The environmental variables Landsat Operational Land Imager (OLI) sensor of P/R: 141/042, *Date of Pass*: 15/10/2015) with a spatial resolution of 30 m and 11 spectral bands with revisit frequency of 16 days. The OLI products used in this study are the renormalized difference vegetation index (e.g., square root of NDVI and DVI (difference vegetation index)), the tasseled-cap transformation (T_{cap}) technique (wetness index, WI), and land use/land cover classification. The WI is generated from OLI image through T_{cap} (Qui et al. 1998). The transformation formula for the OLI scene is defined in Crist and Cicone (1984), and is implemented in the model using model builder tool of ERDAS IMAGINE software (version 9.1, *Atlanta*, Georgia, USA). A LULC map was obtained by performing a supervised maximum-likelihood classification technique (Curran et al. 2000).

6.7.6 Socioeconomic Variables

Socioeconomic variables, like population density, illiteracy rate, nonworking population, and tribal population, play an important role in sustaining transmission of VL and its endemicity. A worse socioeconomic condition is associated with the risk of seroconversion (Picado et al. 2014). The socioeconomic data are collected from the 2011 census. The population density, illiteracy rate, tribal population percent, and nonworking population for each village are calculated individually, and weights for different sub-layers have been assigned accordingly.

The past history of the KA incidence rate is overlaid with the socioeconomic data, and spatial correlation analysis has been performed. Based on the spatial relationship with higher incidence rate, the threshold value for each individual theme has been marked.

6.7.7 Kala-azar Risk Mapping

The methodology used to develop a geo-environmental KA risk map in GIS platform is illustrated in Fig. 6.1. According to the degree of spatial correlation between geographical factors and KA incidence rate, simple weightings/ratings are calculated for all of the input variables. The rating systems were based on values from 1 to 5, where, "5" means highly suitable, "4" highly to moderately suitable, "3" moderately suitable to less suitable, "2" less suitable (unsuitable), and "1" highly unsuitable and kala-azar-restricted area. The scores and weights of the 14 geographical variables are determined based on the KA incidence rate for each category through overlay analysis. Each input raster is weighted a percent influence, with the total influence for all raster equaling 100%. The highest influence of the environmental parameters is considered as 40%, the epidemiological parameter is 20%, the socioeconomic and climatic parameters are considered as 15% for each, and the topographical parameter is considered as 10% each. The output raster is created by the multiplication of cell values by their percentages of influence. Kala-azar risk map is identified based on the following formula:

$$KRM = HR_{cl} \times HR_{t} \times HR_{env} \times HR_{se} \times HR_{epidmology}$$

where KRM is kala-azar risk Map, HR_{cl} is risk of climatic variables, HR_{t} is risk for topography, HR_{env} is risk for environmental parameters, HR_{se} is risk for socioeconomic parameters, and $HR_{epidmology}$ is risk for epidemiological parameters.

6.8 Results and Discussion

The kala-azar risk map for climatic parameters is computed using temperature and RH, topographic parameters (absolute and relative relief), environmental parameters (vegetation, LST, WI, LULC), socioeconomic parameters (population density, illiteracy rate, tribal population, nonworking population) are computed and weights are assigned for each theme individually. The kala-azar risk for epidemiological parameters (incidence rate and disease distribution) shows strong correlation of spatial distribution pattern of it over the study area and is generated through spatial analyst module which provides valuable inputs for KA risk identification. Weights

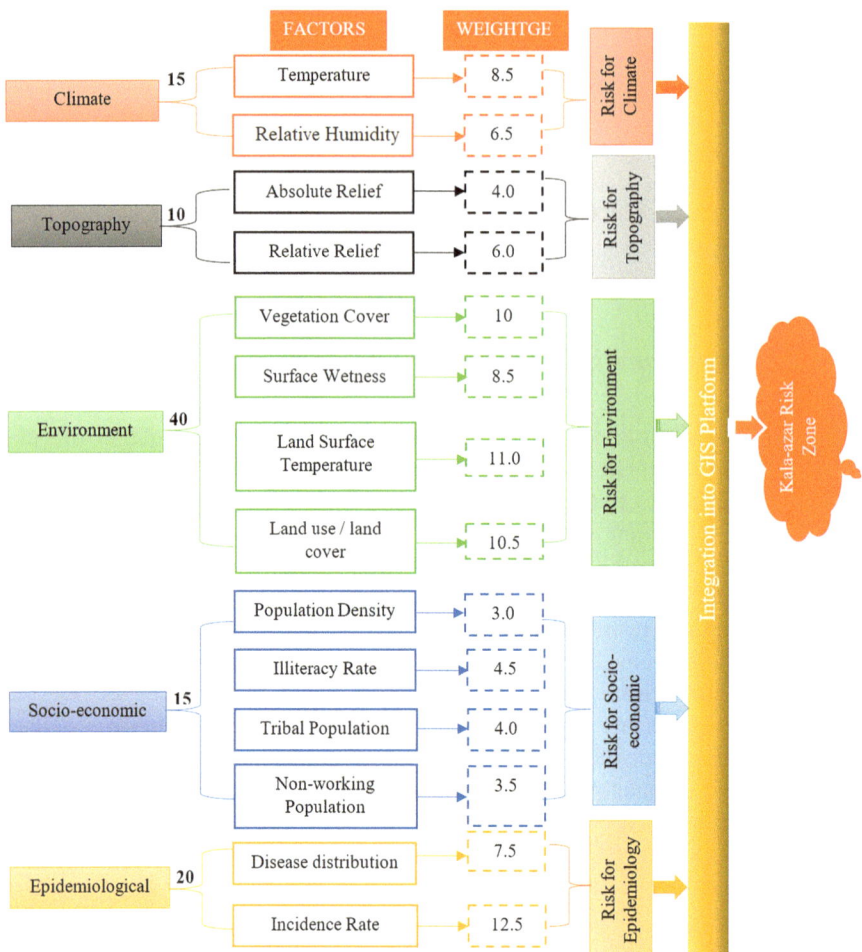

Fig. 6.1 Flowchart map of kala-azar risk mapping

are assigned to these parameters (8.5% temperature, 6.5% RH, 4.0% absolute relief, 6.0% relative relief, 10.0% vegetation, 8.5% LST, 11.0% WI, 10.5% LULC, 3.0% population density, 4.5% illiteracy rate, 4.0% tribal population, 3.5% nonworking population, 12.5% incidence rate, and 7.5 disease distribution) for integrated mapping. The final output raster is classified based on the risk level into five subgroups as very high, high, moderate, low, and very low. The results of the study also showed that 6.8%, 13.3%, 24.6%, 44.3%, and 11.0% of the area are subjected to very high, high, moderate, low, and very low risk, respectively. The very high areas are mainly distributed in the east and south west of the study area. The very high-risk areas are characterized by temperature range of 28.48–29.08 °C, RH varied from 71.09% to 80.60%, absolute relief varied from 43 m to 45 m, relative relief varies from 26 m to 32 m, plantation with settlement, mean RDVI value, low-lying moist area, WI values in the range of 13.97–16.75, LST ranged from 23.07 to 39.27, area dominated by high illiteracy rate, maximum population density, maximum percent of tribal population and agricultural population, disease incidence rate. The high areas are mainly distributed in the west, small pockets of east, and south of the Jandaha block. The very low-risk areas are identified in the central and north of the block (Fig. 6.2). The low-risk areas are characterized by low and extreme temperature areas, low RH, relative relief (less than 22 m), absolute relief (less than 35 m), minimum RDVI value (less than 0.08), maximum RDVI value (more than 1.20), low LST (less than <23 °C), high LST (more than 35°CC), bare surface areas, waterbodies, crop land, high literacy rate, higher number of working population, low tribal population density, and areas with very low incidence rate, almost nil.

Kala-azar is a local and focal disease. The distribution of KA and intensity vary from place to place. Nevertheless, the transmission of KA is enduring unabatedly (simmering transmission) in the entire area and is dependent on micro-level environmental (physiographic and climatic) and demographic variables. Physiographic factors mainly subsidize to sandfly density, while climatic features effect extrinsic incubation (of parasites) unswervingly along with vector survival. The nature and degree of change in the incidence of parasitic disease are exaggerated by amends in land use, changes in vegetation characteristics, changes in local climatic characteristics, and many other environmental factors (Bhunia 2014). The trilateral association of vector, disease outbreak, and geographical factors is accomplishing

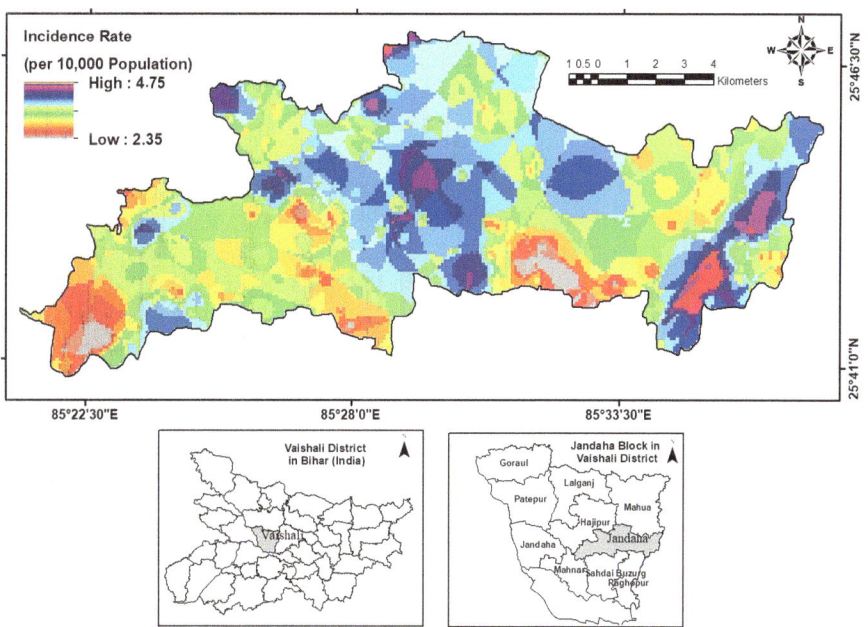

Fig. 6.2 Risk zone map of kala-azar in Jandaha block of Vaishali district (Bihar, India)

importance and prerequisite to address such issues through GIS-based decision support system. In addition to ecological parameters, some significant local aspects such as climatic, socioeconomic, environmental, and topographic patterns of the community play a key role in KA transmission. In this study, plantations associated with settlements, agricultural fallow, and marshy land were considered at higher risk because of their large areal extent in endemic villages. The observed villages are near agricultural lands, yielding grass, or weeds, which offer wetness to the soil in contiguous domestic biotopes of the research sites. Results also show that the low-density vegetation (minimum NDVI) was allied with a high incidence rate of VL as stated by the level of vegetation cover in village neighborhoods (Bhunia et al. 2010a, b). Such database can be used to control a priori sandfly breeding sites and post outbreak, damaging sandfly breeding sites. GIS mapping would generate it easy to up-to-date evidence instantaneously and to recognize the risk areas at micro level that is lowest unit fortified with computer amenities, and the evidence can reach immediately to state and policy makers to verbalize attentive and cost-effective KA control strategy. Hence, these parameters are to be used for decision-making and control of the disease. It can also offer strategies as to which policy could be most appropriate and cost-effective under the prevailing conditions, especially for the developing nations.

6.9 Conclusion

Understanding the epidemiology of leishmaniasis infections and new clinical patterns is useful for disease management and epidemic control. Visceral leishmaniasis becomes an important and opportunistic parasitic disease. Risks of emergence or re-emergence of leishmaniasis are associated with several aspects. The VL is spread naturally in endemic regions and is transmitted to the neighboring areas. Overlap of HIV/VL coinfection is a serious worldwide concern. Hence, HIV and VL programs should develop and combined and well-coordinated intercessions events, mainly in areas where VL is endemic. Association of the two programs should predominantly be reinforced by the local administration at the district level to emphasize on the definite situations of each area of endemicity. The movement of people, for example seasonal migrant laborers, rural-urban migration, emigration activities from areas where disease is nonendemic to those where it is endemic. The existence of immigrants and internal population movements as a result of war and conflicts create key issues for the expansion of both diseases. In ISC, transmission is anthroponotic and VL-HIV-coinfected patients should be embattled as latent reservoirs of infection. Averting the spread of *L. donovani* is obligatory by providing bed nets. In anthroponotic transmission areas, more research is essential to offer enough evidence to endorse secondary prophylaxis action, and the risk of confrontation must also be measured in this circumstance. Moreover, the re-emergence of disease has increased in some areas due to increased coinfection with HIV and tuberculosis in recent decades.

Chapter 7
Measures and Control of Kala-azar

Abstract This chapter describes the various measures and control strategy taken into account for kala-azar transmission. This chapter provides a brief introduction of pandemic history of kala-azar with special reference to World Health Organization and the universal plan adopted for the disease control strategy. A brief overview of disease control strategy has been described in this chapter. The future direction of geoinformatics and environmental and mathematical models to be adopted for disease control programs and strengthening control strategies is discussed. This chapter also describes the proposed approach of telemedicine, smart phone, artificial intelligence, cloud computing, data science, and IoT in kala-azar control programs. A short synopsis of recent and future challenges of disease control strategy is also taken into account.

Keywords Kala-azar control · Satellite communication · Future control strategy · Role of geoinformatics

7.1 Introduction

Spatial distribution of kala-azar/VL expands from the Asian shores of the Pacific Ocean to the passages of Gibraltar. Moreover, the incidence of this disease continues its unabated journey globally (WHO 2010), exacerbated by failure of affordable vaccine development, complemented with an incomplete dose of the drug consumed by the victims owing to lack of awareness of the latter (Das et al. 2010). Declining per capita expenditure on health (Sharma et al. 2006), poverty (Thakur 2000), illiteracy (Adhikari et al. 2010), etc. adds to the menace. Historically, in India, KA broke out in an epidemic form at intervals of

© The Author(s), under exclusive license to Springer Nature Switzerland AG 2020　　　103
G. S. Bhunia, P. K. Shit, *Spatial Mapping and Modelling for Kala-azar Disease*,
SpringerBriefs in Medical Earth Sciences, https://doi.org/10.1007/978-3-030-41227-2_7

15–20 years, each episode lasting 3–4 years. More than a century ago, epidemiologists and other medical scientists instigated to investigate the prospective of maps for understanding the spatial dynamics and intensity of VL, predictive modelling of geographic limits, and its interface with the environment by analyzing the domain acquaintance of geospatial technology. Since a few decades, VL and other vector-borne diseases have resurfaced and spread rapidly. There have been long-term worldwide efforts to eliminate VL, and newly exposed signs of it (e.g., post-kala-azar dermal leishmaniasis (PKDL), HIV-coinfection, etc.) are increasing. Without timely treatment, VL is always fatal. Delays in diagnosis and treatment increase not only the risk of morbidity and mortality but also the risk of transmission of infection to others. An earlier study reported that VL control has been hampered by poorly functioning health services, lack of political will, nonavailability of reliable diagnostic tests, and unresponsiveness to antimonials, as well as inadequate vector control services (Chattopadhyay 2018). On many fronts, inventions are required urgently to control old cases and prevent new ones from spreading. Hence, scientists and researchers in various fields such as biochemistry, genomics, entomology, computing, remote sensing, avionics, artificial intelligence, and robotics have combined their resources to develop new ways to fight the disease. Several efforts were undertaken to strengthen the elimination programme and did not necessarily arise from research needs of the national programmes. With 60% of the total VL encumber in South Asia, the disease has become a public health nuisance in the region (Meheus and Boelaert 2010). VL not only leads to human suffering since unprocessed cases result in death but also has an insightful impact on the livelihood of affected households (Boelaert et al. 2009). Nevertheless, several aspects of the VL burden of disease transmission are still not apparent. One of the major opportunities nowadays for the elimination initiatives of this disease is the advent of geospatial tool and technology which might help us understand the total gamut of the peril and to realize the measures of control through environmental modelling while precisely identifying the VL-risk areas.

7.2   Pandemic History of VL (Kala-azar) and Control Program

In 2015, WHO designated VL as a neglected tropical disease (NTD) due to relatively minimal granted attention from the public, resulting in high mortality rates (Bi et al. 2018). Over the last 50 years, reported national and regional VL incidence has oscillated in an around 15-year cycle (Malaviya 2015). Pigott et al. (2014) collected and summarized reported VL and CL data for the entire world during the period between 1960 and 2012 and listed approximately 55 countries of reported VL cases and 45 countries of unspecified VL cases. Still, thousands of

VL cases may not be considered in the WHO VL assessment report as some countries do not acquiesce their infection report without public health information. Even for the countries with completed public health information system, the documented epidemiological data and official numbers are likely to underrate grossly (Bi et al. 2018). Hence, VL remains a serious and ever-present menace to worldwide public health.

As per the report of WHO, the noteworthy drop in documented VL cases can principally be attributed to a substantial diminution in reported VL cases in India and Bangladesh. From the period between 2006 and 2016, the number of reported VL cases in India declined from 39,173 to 6249 and the informed VL cases in Bangladesh reduced from 9379 to 255. This may be attributed to wide utilization of insecticide-treated nets (Picado et al. 2010b). However, the prevalence of VL did not change significantly for other regions, like Brazil, where annually reported VL cases were approximately 2700 throughout the last decade. The reported VL of Somalia is less than 100 cases in 2006 and is increased up to 781 cases in 2016 (Bi et al. 2018). Hence, if no instantaneous measures are taken in these regions, a large scale of VL epidemic is imminent.

7.3   Universal Plan for VL (Kala-azar) Surveillance

Universal efforts to raise VL surveillance, vector control, diagnoses, and treatment are developing. In 2012, WHO hurled an action plan for dipping the influence of neglected tropical diseases (NTDs), including VL. Currently, WHO took an initiative to describe a road map, recommended by donor partners and stakeholders for prevention, control, elimination, and eradication of 17 neglected tropical diseases, including VL, by 2030, as a step toward realizing the Sustainable Development Goals (Fitzpatrick et al. 2017). WHO has set up a goal to lessen VL incidence to under 1 case/10,000 people/year at subdistrict level in the ISC by 2020.

Critical charities by various stakeholders, including national and international organizations, have buttressed VL eradication efforts in Indian subcontinents, ranging from drug accessibility to sustenance to positioning of intercession from the Bill & Melinda Gates Foundation and many others. Co-ordination with academia, technical and development partners, financial institutions and the pharmaceutical industry is needed to collaborate with regional researchers, national disease control programmes, and policy makers to identify gaps in knowledge, define research needs, and generate evidence to inform the Regional Advisory Group tasked with guiding the regional and national strategy, policy and public health practice for VL elimination. The research area to be focused in kala-azar control strategy is illustrated in Fig. 7.1.

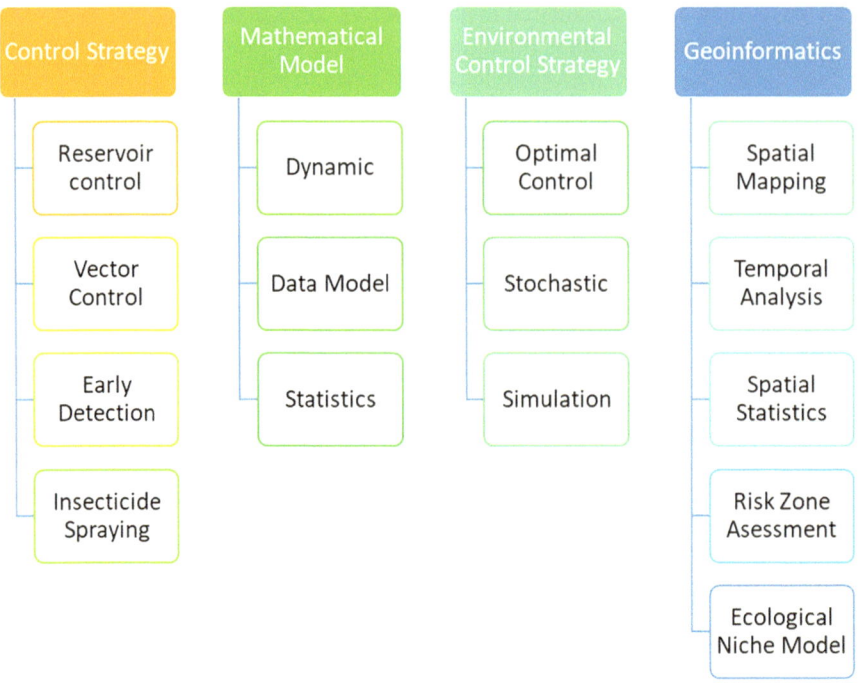

Fig. 7.1 Research tree to be focused for future Kala-azar control strategy. (*Source*: Modified after Bi et al. 2018)

7.3.1   Existing Mitigation Measures of VL (Kala-azar)

Research on VL vaccination was initiated in the 1990s and researchers vexed to employ the proteins from *L. donovani* to progress vaccines (Jaffe et al. 1990). Duthie et al. (2016) studied several different vaccine antigens for VL transmission using recombinant proteins from *E. coli*. Since 1995, researchers focused on ZVL since various ZVL control strategies are related to animals that may be categorized into three main aspects: (a) early detection of human cases, (b) annihilation or treatment of infected dogs, and (c) vector control (Tesh 1995). Guerin et al. (2002) asserted that early diagnosis and treatment are the effective control strategies for VL control. Especially in India, *P. argentipes* is becoming resistant to the insecticide. However, some special challenges such as lack of financial support, remote places, and civil war have been observed. Dantas-Torres and Brandão-Filho (2006) introduced the Brazilian Leishmaniasis Control Program (BLCP) which comprises diagnosis and early treatment of human cases, immunological transmission of seropositive dogs, and insecticide spraying. Quinnell and Courtenay (2009) suggested tropical insecticide-impregnated collars and pour-ons that can be used to lessen VL incidences for dogs more than 83%. Werneck (2014) considered the efficacy of control policies based on the reproduction number and recommended that controls of *L. infantum* density, the ratio of female vector, and the extrinsic incubation period) and controls of dog through culling infected dogs, vaccinations, insecticide releasing. Lengeler (2004) used an alternative ZVL control strategy through indoor residual spraying and insecticide-treated nets (ITNs). Bhunia (2014) suggested suitable eco-environmental factors (e.g., inland waterbodies, surface dampness, soft stem vegetation, low absolute relief, minimum temperature, relative humidity, rainfall, sea level pressure, etc.) for VL propagation and abundance of sandfly. The potential application of this research can provide early warning for kala-azar-risk areas as well as high density of *P. argentipes* for effective insecticide spraying. Picado et al. (2015) summarized the results of the KALANET project to understand ITNs efficacy in India and Nepal and showed 50% reduction of *L. donovani* infections. Özbel et al. (2016) described the geographical dispersal, ecological aspects, and species habitat of VL vectors in Bangladesh.

To predict epidemic transmission using mathematical model has become a recent trend in VL research (Bi et al. 2018). Rapid advancement in computer technology has resulted in computer-aided simulation that helps mathematical model directly predict future VL prevalence. In India, before 2005 there was no national VL case reporting and surveillance system. Active and passive case detection records and statistics were not available, and the information on the exact number of VL cases, geographic distributions, etc. were not noticeable. As per the National Vector Borne Disease Control Program (NVBDCP), VL cases were reported compulsorily in India. The control strategy for VL is based on aggressive identification of cases, effective management, and vector control measures for the reduction of not only morbidity and mortality but also disease transmission. However, substantial research and progressions are still required to attain effective control for humans against VL parasites.

7.4   Future Direction of Geoinformatics in Kala-azar Control

Mostly, RS was used to recognize elements of infectious disease and to create models to envisage their progression. Meteorological factors, topography, vegetation density, ecological disturbances, proximity to waterbodies, presence of hosts, and other socioeconomic factors favor the prevalence of VL transmission and sandfly distribution. Patz et al. (2000) reported that ecological disturbances such as deforestation, migration, road constructions, and El Niño phenomenon are considered to influence the emergence and spread of leishmaniasis. Geoinformatics are used to determine the environmental factors related to the sandfly and VL transmission by integrating spatial and the non-spatial referenced data.

7.4.1 Increasing Extent and Operation of Remote Sensing Data

The altering climatic and ecological circumstances, human behavior, migration, unplanned urbanization effect the periodic and spatial distribution of sandfly, host, and transmission of pathogens. Most of the environmental variables (climatic, ecological, and hydrological) affect the transmission cycle of VL between pathogen agents, vectors, and intermediate hosts that can be mapped proficiently from remote sensing data that convey specific resolution of sensor. Between 2010 and 2019, there has been a noteworthy evolution in EO satellite, including in an around 700 satellite in space, providing improved spectral and spatial resolution with shorter repetitive coverage, empowering improved earth monitoring at universal level (Parselia et al. 2019). Still, the better temporal and spatial resolution of data and data accuracy are the principal barriers. Hence, high-resolution remote sensing data aided to better understand the geographic distribution, abundance, and dynamics of VL transmission and associated vectors and pathogens. The application of novel digital image processing methods such as spatial data mining, support vector machine, neural network analysis, and object-based remote sensing analysis needs to be explored for disease control programs. Additionally, to scale up predictions and move from the local to regional level analysis can be addressed ideally combined with microwave (SAR Sentinel) and hyperspectral (Hyperion, HySI, AVIRIS) satellite data. Chronological data need to be added to RS applications for human health to forecast the disease outbreak. High spatial resolution RS data (Worldview, GeoEye, Cartosat, DigitalGlobe) should be reconnoitered to identify environmental variables to achieve latent paybacks for disease control strategy. Despite the advancement made in epidemic predicting, there is still the need for new prevailing modelling methods like artificial intelligence and collaborative modelling featuring long-lasting satellite observations that permit the determination of extremely multi-faceted associations across data and risk aspects related to the VL transmission.

7.4.2 Mathematical Modelling for Kala-azar Transmission

Mathematical modelling provides tools to appraise interferences to designate both the intensity and timetable over which an intervention might have to be conducted. Hartemink et al. (2011) developed a model by using a range of various assumptions regarding disease progression in humans, the intensity of transmission, and the role of sandflies. Most VL models are "compartmental" models, where the host passes through several phases of the disease at different rates and the assumptions are strictly knotted to the current understanding of the biology. Rutte et al. (2017) used various types of mathematical models, such as (i) Erasnus MC model (symptomatic and asymptomatic individuals are the sole contributors to transmission) and (ii) Warwick model (converts the Markov model of the natural history of VL into a transmission model with vector population dynamics in which asymptomatic individuals are the main contributors to transmission), to understand the transmission of VL in India. Models of VL do not include all of the known complexity of VL biology, mainly because of a lack of data to define the stages, quantify the rates of progression between stages, and chart the time evolving distribution of the infected population between stages.

7.4.2.1 System Dynamic Model

Dye (1996) first presented susceptible-latent-infectious removed (SLIR) ordinary differential equation (ODE) to describe VL epidemic, and the model considered the transitions between these populations. However, in this model, the researcher does not consider the behaviors of dog and vector in this model. Ribas et al. (2013) describe VL transmission among humans, dogs, and vectors, including the susceptible (dog, sandfly, and human), latent (dog, sandfly, and human), infectious (dog, sandfly, and human), and recovered (dog and human) populations. Subsequently, WHO's designation of VL as an NTD in 2015 number of research studies have considered on emerging mathematical model of VL. Zhao et al. (2016) introduced the ODE model to widely define VL epidemic, comprising a hospitalized inhabitant. This population has a greater prospect of endurance than infections short of hospitalization because of the efficient treatment. In this model, researchers applied a regressive bifurcation technique to investigate VL equilibrium behavior and the basic reproduction rate (R_0). Biswas et al. (2017) planned a compartment-based ODE model of VL transmission to elucidate disease transmissions in symptomatic VL, asymptomatic VL, and PKDL-infection classes. Researchers condensed the intricacy of system sensitivity analysis and abridged the figures of expected model parameters. Most of the ODE models designate VL epidemics and transmission. The expansion of a novel dynamic model is an active area of exploration in the study of complex transmission performances of VL under numerous circumstances and the growth of enhanced vindication and control strategies.

7.4.2.2 Statistical Model Based on Real-World Data

Several researchers realized the importance of data utilization in the development of the VL model. The VL statistical model is employed to identify key parameters in VL transmission process and determine relationships between the number of parameters and the number of infected populations. The epidemiological data are principally used in three ways: (a) developing statistical model based on disease incidence data; (b) predicting future prevalence based on historical report; and (c) calibrating model parameters in mathematical epidemic models using existing data. The primary concern of the statistical model is to identify the significant parameters in kala-azar transmission and to determine relationship between the key aspects and number of infected populations. Werneck and Maguire (2002) developed spherical covariance structure model and analyzed VL disease prevalence using census tracts in different regions of Brazil. In 2007, Werneck et al. developed 21 statistical models and found a significant correlation between human infections in residential areas with green vegetation, infected dogs, urbanization index, and socioeconomic status index. Thompson et al. (2002) investigated the relationships between climate and VL epidemics by establishing the statistical regression model. The impact of geographical features of residency areas on VL transmission was also considered. Kesari et al. (2010) studied the association between *P. argentipes* and housing characteristics between endemic and nonendemic district of Kala-azar in India based on multilevel logistic regression analysis and observed that mud plastered wall, mixed dwelling, and area were strongly associated with the presence of vector. Bhunia et al. (2011a) studied the nearest neighbor analysis to the measure the KA distribution pattern. The author also considered Poisson regression analysis to predict sandfly density between the measured distance from the river banks and also took the river types (perennial and nonperennial) into account. Bhunia et al. (2012a) developed a polynomial regression model-based poison distribution of disease incidence data to determine the relation between VL transmission and vector density with the seasonal variation of vegetation characteristics in India. In 2013, Bhunia et al. established the hotspot and coldspot analysis for VL transmission based on spatial statistical approach (Voronoi statistics, inverse distance weighting, Moran's I, Getis-Ord G_i^*) and future direction of disease transmission. Kesari et al. (2013) explored the relationship between the seven explanatory variables (i.e., minimum RDVI, maximum RDVI, mean RDVI, minimum LST, maximum LST, mean LST, and season) and *P. argentipes* density by computing Pearson's correlation coefficient. Structural autoregressive integrated moving average (ARIMA) models were used in order to model the sandfly density and VL cases (Bhunia 2014). In this analysis, cross-correlation was applied to assess the degrees of correlation between various weather variables, *P. argentipes* abundance, and incidence of VL cases over a range of time lags from 0 to 5 months. Several geostatistical models (e.g., inverse distance weighting, radial basis function, semivariograms, Moran's *I*, Getis-Ord G_i^*) have been studied to determine the future direction of disease distribution and prediction of probable KA transmission zone (Bhunia and Shit 2019).

7.4.2.3   Ecological Niche Model

Since 2006, ecological niche modelling (ENM) has been used for VL transmission stemming from genetic algorithm. Nieto et al. (2006) approach an ENM for VL transmission in Brazil using GIS and predict the prevalence risk at high, moderate, and low levels. Similar approaches have been adopted for North America (Colacicco-Mayhugh et al. 2010) and the Middle East regions (Gonz'alez et al. 2010). In 2010, Bhunia et al. developed an index model to identify the eco-environmental factors associated with the risk of human VL (Kala-azar) on the Indo-Gangetic Plain. Bhunia et al. (2012a, b, c) studied the probabilistic approach (Bayesian classifier) and developed a "suitability estimates" of land use/land cover (LULC) characteristics for *P. argentipes* distribution. The author also described the geo-environmental approach for VL transmission in endemic and nonendemic areas of KA. Various microenvironmental parameters (indoor room temperature, indoor relative humidity, vegetation vigor, LULC, location-allocation of surface waterbodies, wetness index) have been considered to delineate the sandfly abundance zone (Bhunia and Shit 2019).

7.4.2.4   Optimal Control Strategy Model

In 1950, Pontryagin developed the optimal control strategy (OCS) which provides optimal measure by maximizing or curtailing a specified objective function depending on limitations identified in differential equation model. Zhao et al. (2016) suggested optimum control into their 12-equation ODE mathematical model, comprising the susceptible, latent, infectious population for sandflies, humans, dogs, recovered human being and hospitalized human population. Agusto and ELmojtaba (2017) considered the use of fabrics and insecticidal animal dollars for the control of VL- and PKDL-infected human populations. Biswas et al. (2017) calculated infection averted ratio (IAR, i.e., proportion of the number of infections prevented to the number of recoveries) and incremental cost-effective ratio (ICER, i.e., supplementary cost per additional health outcomes) and suggested a tactical mechanism, integrating various combinations of optimal strategies for VL and PKDL infections.

7.4.2.5   Control Strategy Using Model Simulation

Simulation is the most common method of mathematical modelling of VL control strategy by considering the epidemic performance of exposed humans, infected humans, and vectors. By changing the model parameters, control levels can be manipulated. In VL transmission, human prevalence is the most sensitive to vector controls. Hence, vector control is the most effective control strategy. Simultaneously, spatial simulation provides spatial information throughout the model behaviors. Bhunia et al. (2013) used annual average VL incidence data to build geo-environmental risk factors for VL transmission. In 2013, Bhunia et al. used past history of kala-azar and predicted the trend and probable zone of VL transmission at the village level through geostatistical modelling in India. The sandfly abundance zone has been predicted by integrating local-level environmental factors into the GIS platform, thereby reflecting high infection density in Vaishali and Muzaffarpur districts in Bihar (Bhunia 2014). Karagiannis-Voules et al. used Brazil's historical VL incidence data (2001–2009) to predict the VL prevalence which reflects the high infection density in 2010.

7.5 Satellite Communication and Kala-azar Disease Control

Satellite communication has the ability to transfer information from one area to another area based on the geostationary satellite. This communication system has two main components: (i) mobile transmission/reception and ancillary equipment and (ii) space segment. The signals are linked between space and ground through uplinking (e.g., transmission of a signal from an earth station to a satellite) and downlinking (e.g., retransmits the signal back to earth). It is performed through mobile receptor, satellite phones, and other handheld devices. Telemedicine is one of the prime applications of satellite communication, in which two parties connect to a third party at a distance. Through telemedicine, healthcare providers in rural areas were connected with the medical experts for consultation and technical support for the diagnosis. Public health centers and hospitals may be connected with the telemedicine network to get access expert opinion by General Practitioners, nurses, paramedics, video conferencing or textual exchange (Dietrich et al. 2018). Use of telemedicine approach has been reported in the kala-azar study by Bhunia et al. (2012a, b, c).

In kala-azar surveillance system, telemedicine approach can be applied in tele-education, *health-on-the-go*, and Satellite for Epidemiology (SAFE) system. In tele-education, awareness of the people about the disease, instruction for the destruction of vector habitats, instructions to rural doctors and health workers and so on. Two-way communication enables diagnosis and treatment of the patients for rural area and remote places. *Health-on-the-go* can provide treatment and transmit health information (text, laboratory test, images, etc.) based on satellite communication. For example, the TraumaStation is a portable medical device which comprises an ultrasound, electrocardiogram, blood pressure, and oxygen meter apparatus. This allows for telehealth with instant messaging and real-time video through satellite communication. Also, HOPE mobile device is a biometric measurement of body mass index, cholesterol, glycosylated hemoglobin and retinal screening from a mobile unit and sending this information to the medical expert's vis SMS/email through satellite for suggestion and treatment of the patients. SAFE is a early health warning system combines with satellite, radio, wireless networks and GIS to promptly recognize and replies to Kala-azar and infectious disease outbreak. These systems may translate to high-quality videos to observe doctors at remote locations in emergency. However, the scientific community of space is continuing their research to offer adequate support for disease surveillance and public health management system. Additionally, new medical research led to the expansion of new medical measures that may be pertinent for the kala-azar disease surveillance and other health-related issues.

7.6   Smart Phone in Kala-azar Control Program

Mobile phones are a vital as well as a popular communication tool in modern society. Mobile technology having a huge potentiality to improve health care and public health services, especially in low- and middle-income country dearth of infrastructure and communication channels and substantial challenges related disease surveillance. Mobile phones and their sensing capabilities have been used as mobile sensors (Fig. 7.2). Sensors are placed statically in case of monitoring patients' conditions, sharing message, tele conversation etc. using telecommunication protocols such as 3G/4G, Wi-Fi, GPRS/UTS or even SMS/MMS messages and Bluetooth, WhatsApp. Mobile apps are used for providing health education information (Fukuoka et al. 2015), self-reporting (Turner-McGrievy et al. 2013), monitoring

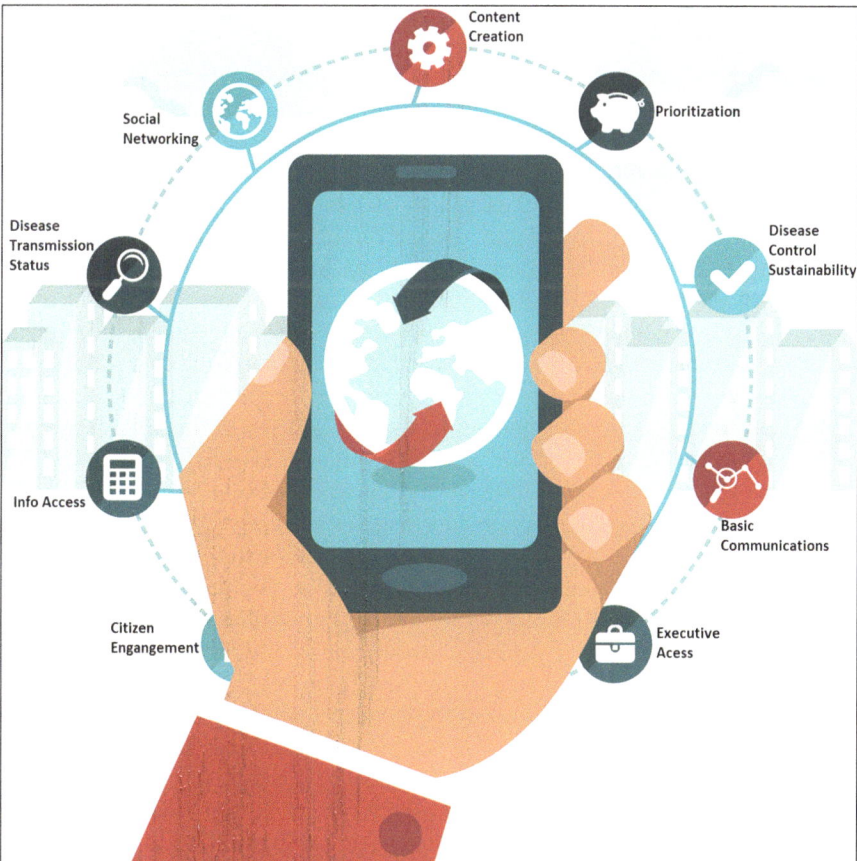

Fig. 7.2 m-health in kala-azar control strategy

(Lee et al. 2018), data collection (Glynn et al. 2014), laboratory test sharing, and submitting feedback (Lee et al. 2018). Subsequently, mobile app programs are used to identify the efficiency of patient's self-care along with the symptom management. Presently medical and nursing interventions using mobile phones and apps have increased. In Kala-azar research, to tailor and personalize care, direct message between frontline workers, programme managers, patients and communities Mobile-phone-based applications (MPBA) are booming rapidly. Hence this can be adopted and implemented in the kala-azar surveillance system to alert people during emergency situations and outbreaks, especially in remote areas, villages, and challenging terrains. Furthermore, smart phones could be used as a tool for strengthening health management information systems to facilitate collection and compilation of information from wide areas.

7.7   Artificial Intelligence in Kala-azar Control Program and Strategy

Artificial intelligence (AI) is an emerging field in the VBD control program which can solve issues (e.g., object detection, visual interpretation, mathematical programming) intelligently difficult for human beings but comparatively easy for programmable computers. Entomological characterization and identification of sandfly species is essential to map and organize the control measurements by epidemiologist where transmission is vigorously occurring. Machine learning is a part of AI that may be used to KA control programs by extracting information from existing data. For instance, intelligent sandfly's trap can be designed with the functionality to classify female *Phlebotomine* species and kill them or to attract the female sandfly to lay eggs in the trap. This classification process can be done through laser sensor and/or audio analysis technique of various species of sandfly recognition. Whenever a new insect approaches the trap, it will mechanically categorize and take decision – release it or kill it. Subsequently, automatic sandfly species classification using infrared recording device for sketching the wingbeat of the in-flight sandfly species. Another issue is the damage of the sandfly's body which can affect the evaluation of the morphological characteristics. Ouyang et al. (2015) used the machine learning model for classifying the mosquitoes based on their wingbeats and to attain more than 80% accuracy. Another possibility for the identification of vector species using DNA barcodes derived through molecular techniques that allows to estimate enormous number of vector data, obtaining online information of population density and the correlation between the incidence rate and mortality du to VL. Kumar et al. (2007) used DNA barcode for identification of mosquitoes in India. Lorenz et al. (2015) used AI approach for the classification of mosquito species with an accuracy of 86–100%. Through AI, it is possible to acquire and store the behavior of sandfly and correlate data such as time, season of high intensity, species captured, and indoor climatic data (temperature and humidity). Subsequently,

AI also support community participation (e.g., identification of sandlfy by visual examination from human trained technician) to improve the performance of a human care system. Machine learning and deep learning techniques are used to classify the different stages of the sandfly's life cycle – eggs, larval, pupae, and adult. The support vector machine (SVM) in digital image processing may be used for visual identification for sandfly morphology. A camera will be unified with a circuit board, where images are fed to an SVM, matching the sandfly's body characteristics.

However, the applications of machine learning and deep learning techniques increase day by day due to the flexibility of their algorithms, but there are still some bottlenecks to overcome. For the implementation of AI, historical data are required for algorithm learning and to predict a reliable outcome. The accessibility, clearance, and changeability of these existing data are decisive for the computer learning process. Subsequently, very small size of dataset are obtainable in open platform, which shots to be tough to acclimatize the model and resolve the problem with proper precision and reliability.

7.8 Cloud Computing in Kala-azar Control Strategy

Presently, cloud computing (CC) architype has become one of the most recent topics in information technology due to its mobility, scalability, and security paybacks by providing on-demand calculating resources such as hardware, storage, services, servers, networks, and applications (Dang et al. 2019). CC is accountable for storage, sharing information infrastructure applications, providing automatic subscription, and processing the data which cannot be processed by fog layer (e.g., decentralized computing setup in which data, calculate, storage and applications are situated someplace between the data source and the cloud). These layers consist of four components, such as cloud storage, information fortification, disease outbreak, and health communication (Fig. 7.3). CC comprises of massive amounts of storage to hoard investigation consequences and accumulated health related data of each user and strongly share among official users, medical practitioners, dispensaries, public health centers and healthcare specialists. Emergency and diagnostic cognizant messages produced from fog layer are also stowed on CC for supplementary investigation by professionals to take instantaneous action and offer precautions in case of emergency.

Through CC, health data can be shared with health professionals, caregivers, and patients in a more structured and organized way. In VL control strategy, tenant database and shared database are implemented to access disease data in various control modules. After that the healthcare data annotation solves the data heterogeneity issue and integrates various data in a patient-centric pattern for cloud applications. The data related to KA/VL are transmitted to CC and are described as follows:

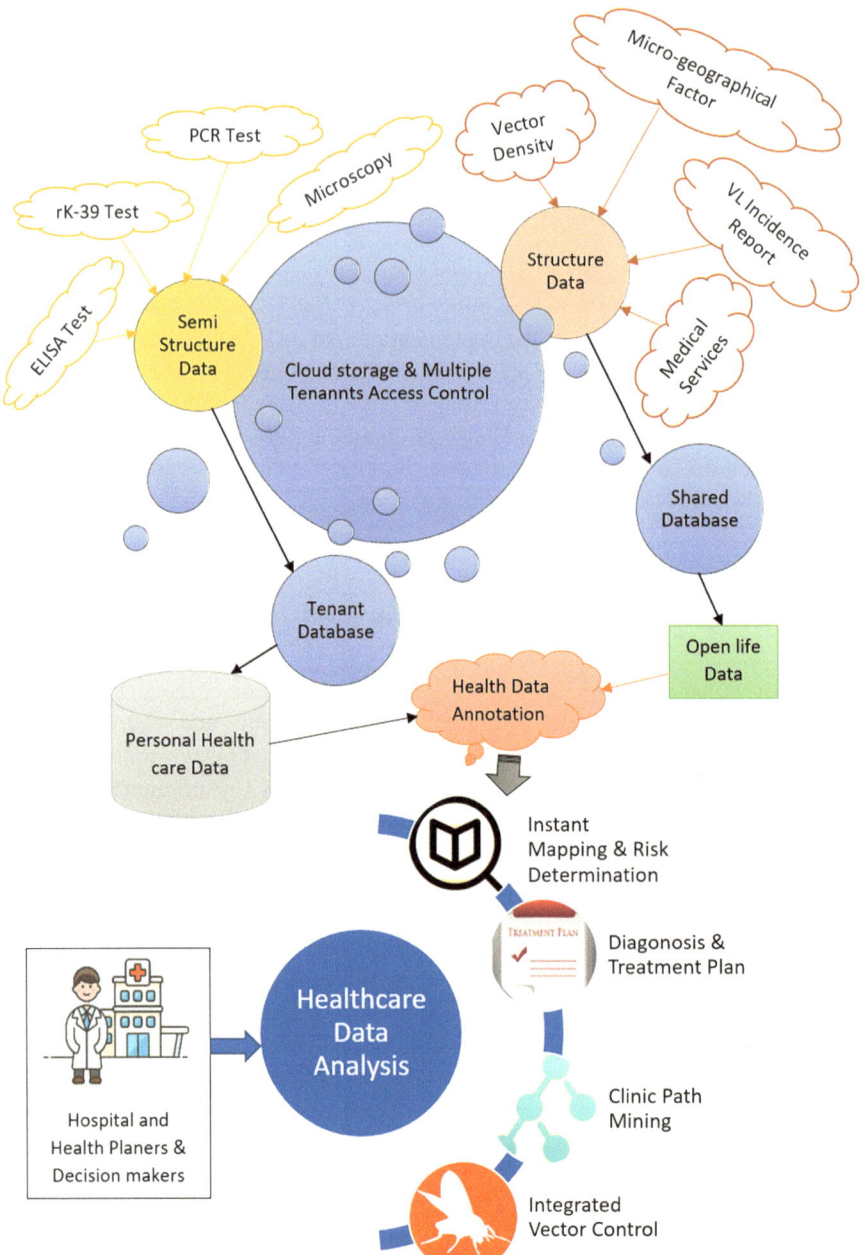

Fig. 7.3 Functional platform of cloud computing-based VL control program

(i) Health data – comprise vital signs of VL incidence. These include high fever, body pain, weight loss, splenomegaly, anemia, etc.

(ii) Environmental data – entomological data such as sandfly density, man-hour density, breeding sites, species distribution, etc. to be collected through sensor. Indoor room temperature, indoor relative humidity, rainfall, carbon dioxide around human dwellings, and other peridomestic environments are also measured by the sensor to evaluate environmental condition under which sandfly lay eggs.

(iii) Medicinal data – include the medication procedure of the patients and is comprised of the medicine name, medicine form, quality, ingestion time of medicine, etc. These data may be acquired through RFID tags.

(iv) Location data – include patients' locations, PKDL cases, HIV/VL coinfection, contact information of suspected patients, insecticide spraying area, etc., which are collected through GNSS sensors to get the travel history and trend of disease pattern analysis.

(v) Meteorological data – include temperature, rainfall, humidity, sea-level pressure, wind velocity, etc., which are collected through climate sensors.

Despite all the benefits of CC, control and data protection are among the most important challenges that have to be considered in the development of the cloud network (Mehraeen et al. 2017). Security is one of the main obstacles to the growth of CC in the health field as the necessity for higher level of data amalgamation, interoperability and allocation among various healthcare physicians and organizations. Hence, public health centers should be able to follow standard strategies and recognize security in healthcare CC. Finally, the health-care data analysis system analyzed cloud storage data to assist clinical decision making, integrated vector control, risk determination, clinical path mining, etc. (Fig. 7.2). Generally, cloud data centers are centralized geographically and situated far away from the end-users.

Hence, CC can be implemented to serve a variety of demands for kala-azar disease control by sharing information to observe results and evaluate information from the end-users that require immediate real-time feedback, remote monitoring, etc.

7.9   Data Science in Kala-azar Control

Internet-based surveillance method for communicable diseases was first introduced in the mid-1990s, when ProMed system was introduced to solicit via email or other media (Madoff 2004). The range of Internet data sources included Google, Wikipedia, Twitter, Internet newswires, and other search engines that enhance the disease surveillance system. Facebook, the third most visited website in the world, is usually less acquiescent to develop communicable disease surveillance due to the dearth of public access (Pollett et al. 2017). Subsequently, automatic data extraction and content analysis from YouTube videos are also exciting. As the VL affect mostly low- and middle-income regions of the

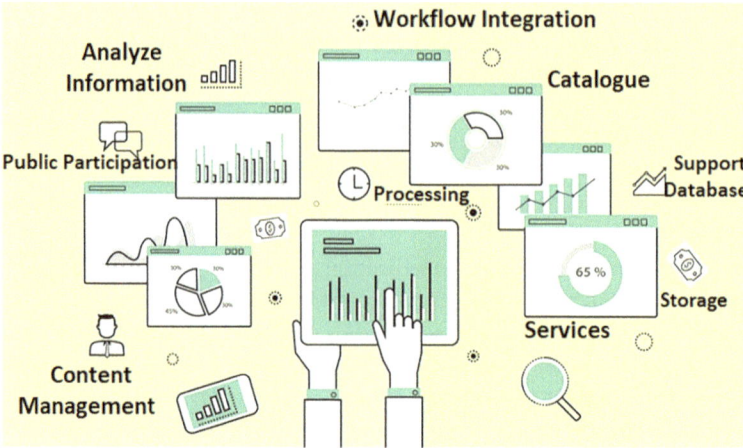

Fig. 7.4 Data science in kala-azar control program

world, due to limited conventional surveillance infrastructure and the data distributed in delay to key stakeholders. Internet-based bio-surveillance system may thus be useful as a supplementary, timely surveillance signal to aid in VL preparedness, situation awareness, and response to disease control program. With the development of IoT and big data analytics in public health, data can be collected from remote areas where the collection has been done manually through government data repository or not done at all. Since, data science is a network of interconnected system, without the necessity for the wider internet, in inaccessible areas to assemble data. These data can finally be allied to the wider grid to assimilate it with the worldwide public health management systems to not only track diseases in real time but also to smear prophetic analytics to avert their spread (Fig. 7.4). In VL (KA) control, the deficiency of readily accessible and germane data mandatory to test hypothesis has run to nonexistence of evidenced-based tactics. Technology and the beginning of data science can straightforwardly overawe this challenge. In an emergency outbreak, an evidenced-based analysis can be better equipped by accumulating data from remote locations and feeding it into the global health system along with data from other sources. Moreover, the precautionary measures can be recommended using the data science to further regulate if the controlling measures proposed are being executed suitably. However, the role of data science associated public health and in averting the feast of vector-borne diseases grow regularly as well as related technology. Hence, the proper planning and careful operation is essential using the proper technology platform and tools.

7.10   IoT in Kala-azar Control Strategy

The Internet of Things (IoT) constitutes the interlinking of *things* with the network connectivity that processes data from the sensor node to the end point. Presently, Information and Communication Technologies (ICTs), particularly IoT-enabling technologies and devices, like smart objects, connected sensors and actuators, wearable sensors, mobile devices, and so on, automatically process and send the required information such as geographic area, patient's electronic medical record (EMR), set of symptoms, laboratory test, outbreaks, day-to-day case report, real-time information of disease patterns and proposed expert suggestion, etc. to the backbone network for smart kala-azar surveillance system. The main component of IoT is based on Radio Frequency identification (RFID) tags, barcodes and QR code that can automatically identify constant or moving entities to monitor and control objects via internet. IoT based KA surveillance system will be helpful by providing information trend of geographical distribution of disease, attain information of timely and accurate survey and insecticide spraying, timely warn to the citizen, check the stocks of medicine, beds in hospital, availability of doctor, by avoiding of underreporting of the cases, and the decision making for possible solutions. A general architecture of IoT-enabled system for kala-azar control program is illustrated in Fig. 7.5.

The sensing layer is designated to collect entomological and epidemiological information, camera for patient's information, the GNSS/GPS sensor for positioning and localization, RFID for identification and smart phones to be used for sensing the information of user's immediate surroundings, such as the home condition, peridomestic environment, sharing of symptoms etc. Secondly, the network layers enable the efficient and secure data transmission to corresponding data processing units, for example, 6LowPAN, NBIOT, LoRa communication protocol to be applied. The data processing module is responsible for retrieving valuable knowledge from the sensing layer. Finally based on the above condition, intelligent services and applications can be provided, such as early warning, disease diagnosis, behavior recognition, and smart support. Hence, IoT permits real-time forewarning, pursuing, and monitoring which allow hands-on behaviors, better precision, and apt intercession by doctors and expand comprehensive patient care delivery more proactively than ever before.

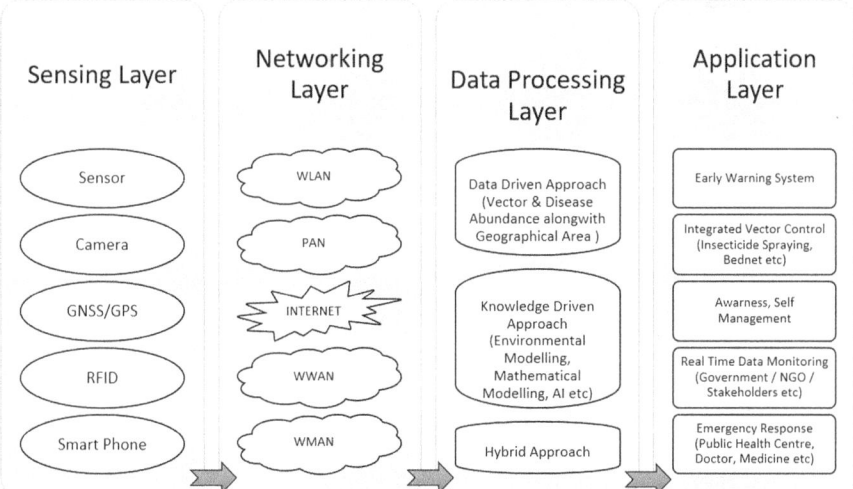

Fig. 7.5 Overall framework of embedding IoT technologies for Kala-azar control program

7.11   Geoinformatics and IoT in Kala-azar Control

In geoinformatics, basic logical processes include buffering, overlays, neighborhood-based analysis, and map algebra. In the IoT scope, mostly buffering and overlaying have been documented. For example, overlay in GIS procedure that place over multiple data sets along with the purpose of recognizing relations between them. A combination of buffering and overlays have been used in assessing health hazard and risk analysis, calculating landscape variables and climatic parameters, and spatiotemporal pattern analysis for disease control program. In geospatial IoT research, spatial analysis of point patterns, cluster, kernal and density analysis and estimation have been found. This will help in examining the spatial relationship of the locations of the measured quantities and known quantities of some phenomenon. Additionally, risk maps are commonly used in mobile for tracking applications, geotagging for visualization, identifying disease phenomenon, and spatial data mining. Besides this geolocational information, various sources of data may be used for enriching the geospatial analysis. For example, this data may be used for web-based sources such as online news, validation official reports, unofficial online report, road data, medical facility, information, demographics, and other optional information of disease risk pattern.

However, challenges of geospatial analysis in IoT projects include the cost of IoT equipment, data storage needs, the requirement events processing and computational analysis of geospatial big data, and the use of databases that natively support spatial data types. Additional open research challenges include the nonappearance of accuracy on simplifications, particularly interpolation, the necessity for standardizations of spatial data infrastructures, and the presence of semantic web concepts to sensor detection, where the web could become an information system where location-related information can be more easily shared across different applications (Kamilaris and Ostermann 2018).

7.12   Issues and Challenges

- Dealing with missing data is an issue towards precise location of disease events.
- Calibration of the sensors is important in several cases. The problem with faulty readings is that they are sometimes hard to detect, if the ground-truth data are not available to compare with.
- Some reliability problems occurred during the installation of the sensors at the field. Additionally, GNSS receiver used in experience low accuracy and week GNSS signals in high density of buildings/vegetative covers areas, and suggested a tag location principle technique to improve this accuracy.
- Accuracy issues may also appear in remote sensing data. For instance, satellite sensor used recording surface temperature averaged over a 1 km × 1 km area with an estimated Root Mean Square Error (RMSE) error of 2–4 °C.

- The aspect of security is particularly important in participatory sensing, since access to IoT device is sometimes difficult due to security restrictions of local domains (e.g., firewalls, use of NAT on mobile network routers).
- Privacy of IoT device and their data is also a sensitive issue. Multiple privacy threats have been reported in mobile RFID applications and mobile location-based services.

7.13 Conclusion

This book has investigated the environmental variables that influence the suitable vector habitat and thus transmission of visceral leishmaniasis or kala-azar. In order to establish easily interpretable relationships (between environmental variables and vector density), the local climate and synoptic environmental characteristics (land use/land cover, vegetation, inland water bodies, surface dampness, land surface temperature, indoor climate) are first assessed. Correlative analysis between the environmental variables and VL incidence vis-à-vis sandfly proliferation has been analyzed. The architecture of the integration process of IoT and CC in kala-azar healthcare policy has been summarized briefly.

The present study affirms that geoinformatics was successfully used with geo-environmental and eco-environmental risk modelling methods to provide an estimate of the risk of kala-azar or VL transmission. The risk maps generated in this study reveal a baseline assessment of kala-azar "risk" and "nonrisk" areas. Finally, it would also give a priceless information capability to other related domains, such as bio-surveillance, health care forecasting, and national issues for country stability. Monitoring of vector resistance to insecticides should be carried out periodically from different geographical zones for planning suitable intervention measures.

Glossary

Asymptomatic Active Kala-azar (VL) patients who can be infective to sandflies but not displaying outward symptoms of KA/VL infection.

Cutaneous leishmaniasis Most common form of leishmaniasis and cause ulcers on exposed parts of the body, leaving lifelong scars and serious disability.

Deterministic model Model based on the average behaviors of population and providing the same outcomes for a set of parameters in every simulation method.

Dormant The patients in between KA and post-Kala-azar dermal leishmaniasis (PKDL) having no symptoms but still harboring leishmania parasites.

Exposed Patients having the leishmania parasites, but are not yet infective to sandflies.

Geo-reference To assign co-ordinates from a reference system, such as latitude and longitude.

Kala-azar Acute form of VL, literally translated from *black fever*. KA patients display symptoms such as fever, weight loss, anemia, and enlargement of spleen and liver.

Layer It is a set of geographic features along with its attribute table or an image. Layers have properties such as layer name, symbology, and label placement.

Mucocutaneous leishmaniasis This causes partial or total destruction of mucous membranes of the nose, mouth, and throat.

Non-symptomatic Patients with VL infection, but no symptoms: includes exposed, asymptomatic, and dormant.

Post Kala-azar dermal leishmaniasis (PKDL) PKDL is characterized by a nodular of popular skin rush. It is not life threatening, but reservoir of leishmania parasite.

rK39 Test Rapid dipstick test (rK39 test) is a serological diagnosis test for VL. It is a 39-amino acid repeat that is part of the kinesin-associated protein in *Leishmania chagasi* and is preserved within the *Leishmania donovani* complex.

G. S. Bhunia, P. K. Shit, *Spatial Mapping and Modelling for Kala-azar Disease*, SpringerBriefs in Medical Earth Sciences, https://doi.org/10.1007/978-3-030-41227-2

Seroconversion Seroconversion occurs in 0–10% of nonimmune HCWs who sustain needlesticks from a source case with hepatitis C.

Spatial Relating to space.

Stochastic model This determines the probability of events such as elimination or re-occurrence to be found. Simulated disease dynamics vary every time.

Symptomatic VL Symptomatic infected individuals display symptoms within a defined period of parasite incubation within them.

Visceral leishmaniasis The disease is a chronic and frequently lethal disease caused by *Leishmania donovani* and *Leishmania infantum* somewhere else. VL patients harboring parasite with or without symptoms.

Zoonotic transmission Transmission of VL that affects animals is largely associated with *L. infantum*. *L. donovani* has been found in animals in East Africa.

References

Aagaard-Hansen J, Nombela N, Alvar J (2010) Population movement: a key factor in the epidemiology of neglected tropical diseases. Tropical Med Int Health 15:1281–1288

Abdullah AU, Dewan A, Shogib RI, Rahman M, Hossain F (2017) Environmental factors associated with the distribution of visceral leishmaniasis in endemic areas of Bangladesh: modelling the ecological niche. Trop Med Health 45:13

Addy M, Mitra AK, Ghosh KK, Hati AK (1983) Host preference of *Phlebotomus argentipes* in different biotopes. Trop Geogr Med 35(4):343–345

Adhikari SR, Maskay NM, Sharma BP (2009) Paying for hospital-based care of kala-azar in Nepal: assessing catastrophic, impoverishment and economic consequences. Health Policy Plan 24:129–139

Adhikari SR, Supakankunti S, Khan MM (2010) Kala-azar in Nepal: estimating the effects of socioeconomic factors on disease incidence. Kathmandu Univ Med J 8:73–79

Agusto FB, ELmojtaba IM (2017) Optimal control and cost-effective analysis of malaria/visceral leishmaniasis coinfection. PLoS One 12(2):e0171102, https://doi.org/10.1371/journal.pone.0171102

Ahammed A, Yusuf A, Feroz S, Selim S, Bhattacharyya B, Ahammed I, Rahman R (2016) Household and environmental risk factors for Kala-azar: a case-control study in Tertiary Care Hospital of Bangladesh. J Sci Found 14(2):56–61

Albuquerque PLMM, da Silva Júnior GB, Freire CCF, de Castro Oliveira SB, Almeida DM, da Silva HF, do Socorro Cavalcante M, de Queiroz Sousa A (2009) Urbanization of visceral leishmaniasis (kala-azar) in Fortaleza, Ceará, Brazil. Rev Panam Salud Publica/Pan Am J Public Health 26(4):330–333

Alemayehu B, Alemayehu M (2017) Leishmaniasis: a review on parasite, vector and reservoir host. Health Sci J 11(4):519

Al-Masum MA, Evans DA, Minter DM, El-Harith A (1995) Visceral leishmaniasis in Bangladesh: the value of DAT as a diagnostic tool. Trans R Soc Trop Med Hyg 89 (2):185–186

Alvar J, Yactayo S, Bern C (2006) Leishmaniasis and poverty. Trends in Parasitology 22:552–557

Alvar J, Aparicio P, Aseffa A, Den Boer M, Cañavate C, Dedet JP et al (2008) The relationship between leishmaniasis and AIDS: the second 10 years. Clin Microbiol Rev 21:334–359

Alvar J, Ve'lez ID, Bern C, Herrero M, Desjeux P et al (2012) WHO Leishmaniasis Control Team. Leishmaniasis worldwide and global estimates of its incidence. PLoS One 7:e35671

Andrade-Narvaez FJ, Canto Lara SB, van Wynsberghe NR et al (2003) Seasonal transmission of Leishmania (Leishmania) mexicana in the state of Campeche, Yucatan Peninsula, Mexico. Mem Inst Oswaldo Cruz 98(8):995–998

© The Author(s), under exclusive license to Springer Nature Switzerland AG 2020
G. S. Bhunia, P. K. Shit, *Spatial Mapping and Modelling for Kala-azar Disease*,
SpringerBriefs in Medical Earth Sciences, https://doi.org/10.1007/978-3-030-41227-2

Aversi-Ferreira RAGMF, Galvão JD, da Silva SF, Cavalcante GF, da Silva EV, Bhatia-Dey N, Aversi-Ferreira TA (2015) Geographical and environmental variables of leishmaniasis transmission. INTECH:105–123. https://doi.org/10.5772/57546

Awati PR (1922) Survey of biting insects of Assam with reference to kala-azar for the whole year from November 1921 to October 1922. Biting insects found in dwelling-houses. Indian J Med Res 10:579–591

Badaró R, Jones TC, Cerf BJ, Sampaio D, Carvalho EM, Rocha H, Teixeira R, Johnson WD Jr (1986) A perspective study of Visceral Leishmaniasis in an endemic area of Brazil. J Infect Dis 154(4):639–649

Bavia ME, Carneiro DD, Gurgel Hda C, Madureira Filho C, Barbosa MG (2005) Remote sensing and geographic information systems and risk of American visceral leishmaniasis in Bahia, Brazil. Parassitologia (Rome) 47:165–169

Beck LR, Lobitz BM, Wood BL (2000) Remote sensing and human health: new sensors and new opportunities. Emerg Infect Dis 6(3):217–226

Belen A, Alten B (2005) Variation in life table characteristics among populations of *Phlebotomus papatasi* at different altitudes. J Vector Ecol 31:35–44

Bern C, Hightower AW, Chowdhury R, Ali M, Amann J, Wagatsuma Y, Haque R, Kurkjian K, Vaz LE, Begum M, Akter T, Cetre-Sossah CB, Ahluwalia IB, Dotson E, Evan Secor W, Breiman RF, Maguire JH (2005) Risk factors for Kala-azar in Bangladesh. Emerg Infect Dis 11(5):655–662

Bern C, Courtenay O, Alvar J (2010) Of cattle, sand flies and men: a systematic review of risk factor analyses for South Asian visceral leishmaniasis and implications for elimination. PLoS Negl Trop Dis 4:e599

Bhatt S, Gething PW, Brady OJ, Messina JP, Farlow AW, Moyes CL et al (2013) The global distribution and burden of dengue. Nature 496:504–507

Bhunia GS (2014) An appraisal of environmental determinants of the disease visceral leishmaniasis (Kala-azar) using remote sensing and GIS techniques: case studies of Vaishali and Muzaffarpur districts, Bihar. PhD thesis, Submitted to the Department of Geography, University of Calcutta, Kolkata, West Bengal, India

Bhunia GS, Shit PK (2019) Geospatial analysis of public health. Springer Nature Switzerland AG 2019, ISBN 978-3-030-01679-1

Bhunia GS, Kesari S, Jeyaram A, Kumar V, Das P (2010a) Influence of topography on the endemicity of Kala-azar: a study based on remote sensing and geographical information system. Geospat Health 4(2):155–165

Bhunia GS, Kumar V, Kumar AJ, Das P, Kesari S (2010b) The use of remote sensing in the identification of the eco-environmental factors associated with the risk of human visceral leishmaniasis (kala-azar) on the Gangetic plain, in north-eastern India. Ann Trop Med Parasitol 104(1):35–53

Bhunia GS, Kesari S, Chatterjee N, Pal DK, Kumar V, Ranjan A, Das P (2011a) Incidence of visceral leishmaniasis in the Vaishali district of Bihar, India: spatial patterns and role of inland surface water bodies. Geospat Health 5:205–215

Bhunia GS, Dikhit MR, Kesari S, Sahoo GC, Das P (2011b) Role of remote sensing, geographical information system (GIS) and bioinformatics in kala-azar epidemiology. J Biomed Res 25(6):373–384. https://doi.org/10.1016/S1674-8301(11)60050-X

Bhunia GS, Kesari S, Chatterjee N, Kumar V, Das P (2012a) Localization of kala-azar in the endemic region of Bihar, India based on land use/land cover assessment at different scales. Geospat Health 6(2):177–193

Bhunia GS, Kesari S, Chatterjee N, Kumar V, Das P (2012b) Telehealth: a perspective approach for visceral leishmaniasis (kala-azar) control in India. Pathog Glob Health 106(3):1–9

Bhunia GS, Chatterjee N, Kumar V, Siddiqui NA, Mandal R, Das P, Kesari S (2012c) Delimitation of kala-azar risk areas in the district of Vaishali in Bihar (India) using a geo-environmental approach. Mem Inst Oswaldo Cruz 107(5):609–620

Bhunia GS, Kesari S, Chatterjee N, Kumar V, Das P (2013) The Burden of Visceral Leishmaniasis in India: challenges in using remote sensing and GIS to understand and control. ISRN Infectious Diseases 2013:675846, 14 pages. Available at: https://doi.org/10.5402/2013/675846

Bi K, Chen Y, Zhao S, Kuang Y, Wu CJ (2018) Current visceral leishmaniasis research: a research review to inspire future study. Biomed Res Int 9872095:1–13

Bill R (1999) Grundlagen der Geo-Informationssysteme. Band 1 (Hardware, Software und Daten). 2. Aufl. Wichmann Verlag, Heidelberg, 4th edn, p 454

Biswas S, Subramanian A, EL Mojtaba IM, Chattopadhyay J, Sarkar RR (2017) Optimal combinations of control strategies and cost-effective analysis for visceral leishmaniasis disease transmission. PLoS One 12(2):e0172465

Boelaert M, Meheus F, Sanchez A, Singh SP, Vanlerberghe V, Picado A et al (2009) The poorest of the poor: a poverty appraisal of households affected by visceral leishmaniasis in Bihar, India. Trop Med Int Health 14:639–644

Bora D (1999) Epidemiology of visceral leishmaniasis in India. Natl Med J India 12(2):62–68

Brahmachari UN (1928) Treatise on Kala-Azar. John Bale, Sons & Danielsson Ltd., London

Burattini MN, Coutinho FAB, Lopez LF, Massad E (1998) Modelling the dynamics of leishmaniasis considering human, animal host and vector populations. J Biol Syst 6:337–356

Campbell-Lendrum D, Dujardin JP, Martinez E, Feliciangeli MD, Perez JE, de Silans LNMP, Desjeux P (2002) Domestic and peridomestic transmission of American cutaneous leishmaniasis: changing epidemiological patterns present new control opportunities. Mem Inst Oswaldo Cruz, Rio de Janeiro 96(2):159–162

Carreira JCA, Magalhães MAFM, da Silva AVM (2015) Chapter 6: The geospatial approach on eco-epidemiological studies of Leishmaniasis. INTECH. https://doi.org/10.5772/57210

Chakrabarti S, Sarkar S, Goswami BK, Sarkar N, Das S (2013) Clinico-haematological profile of visceral leishmaniasis in immunocompetent patients. Southeast Asian J Trop Med Public Health 44(2):143–149

Chapman LAC, Jewell CP, Spencer SEF, Pellis L, Datta S, Chowdhury R et al (2018) The role of case proximity in transmission of visceral leishmaniasis in a highly endemic village in Bangladesh. PLoS Negl Trop Dis 12(10):e0006453

Chappuis F, Sundar S, Hailu A, Ghalib H, Rijal S, Peeling RW, Alvar J, Boelaert M (2007) Visceral leishmaniasis: what are the needs for diagnosis, treatment and control? Nature Reviews (Microbiology) 5:873–882.

Chattopadhyay S (2018) An innovative approach for mosquito borne diseases control: an original concept. Med J DY Patil Vidyapeeth 11:232–236

Cheghabaleki ZZ, Yarahmadi D, Karampour M, Shamsipour A (2019) Spatial dynamics of a phlebotomine sand flies population in response to climatic conditions in Bushehr Province of Iran. Ann Glob Health 85(1):60

Chowdhury R, Huda MM, Kumar V, Das P, Joshi AB, Banjara MR, et al (2010) The Indian and Nepalese programmes of indoor residual spraying for the elimination of visceral leishmaniasis: performance and effectiveness. Ann Trop Med Parasitol 105(1):31–45

Chowdhury R, Dotson E, Blackstock AJ, McClintock S, Maheswary NP, Faria S et al (2011) Comparison of insecticide-treated nets and indoor residual spraying to control the vector of visceral leishmaniasis in Mymensingh District, Bangladesh. Am J Trop Med Hyg 84(5):662–667

Chowdhury R, Kumar V, Mondal D, Das ML, Das P, Dash AP, Kroeger A (2016) Implication of vector characteristics of Phlebotomus argentipes in the kala-azar elimination programme in the Indian sub-continent. Pathog Glob Health 110(3):87–96. https://doi.org/10.1080/20477724. 2016.1180775

Clarke K, McLafferty S, Tempalski B (1996) On epidemiology and geographic information systems: a review and discussion of future directions. Emerg Infect Dis 2:85–92

Clements ACA, Lwambo NJS, Blair L, Nyandindi U, Kaatano G, Kinung'hi S, Webster JP, Fenwick A, Brooker S (2006) Bayesian spatial analysis and disease mapping: tools to enhance planning and implementation of a schistosomiasis control programme in Tanzania. Trop Med Int Health 11(4):490–503

Cline BL (1970) New eyes for epidemiologists: aerial photography and other remote sensing techniques. Am J Epidemiol 92:85–89

Cohen C, Corazza F, De Mol P, Brasseur D (1991) Leishmaniasis acquired in Belgium. Lancet 338:128

Colacicco-Mayhugh MG, Masuoka PM, Grieco JP (2010) Ecological niche model of *Phlebotomus alexandri* and *P. papatasi* (Diptera: Psychodidae) in the Middle East. Int J Health Geogr 9:2. https://doi.org/10.1186/1476-072X-9-2

Costa CH, Werneck GL, Rodriques L Jr et al (2005) Household structure and urban services: neglected targets in the control of visceral leishmaniasis. Ann Trop Med Parasitol 99:229–236

Crist EP, Cicone RC (1984) Application of Tasseled Cap concept to simulated Thematic Mapper data. Photogrammetric Engineering & Remote Sensing 50:343–352

Croft SL, Sundar S, Fairlamb AH (2006) Drug resistance in leishmaniasis. Clin Microbiol Rev 19:111–126

Cross ER, Newcomb WW, Tucker CJ (1996) Use of weather data and remote sensing to predict the seasonal distribution of *Phlebotomus papatasi* in southwestern Asia. Am J Trop Med Hyg 54:530–536

Cruz I, Morales MA, Noguer I, Rodriguez A, Alvar J (2002) Leishmania in discarded syringes from intravenous drug users. Lancet 359:1124–1125

Cunze S, Kochmann J, Koch LK, Hasselmann KJQ, Klimpel S (2019) Leishmaniasis in Eurasia and Africa: geographical distribution of vector species and pathogens. R Soc Open Sci 6:190334

Curran PJ, Atkinson PM, Foody GM, Milton EJ (2000) Linking remote sensing, land cover and disease. Adv Parasitol 47:37–80

da Costa SM, Cordeiro JLP, Rangel EF (2018) Environmental suitability for Lutzomyia (Nyssomyia) whitmani (Diptera: Psychodidae: Phlebotominae) and the occurrence of American cutaneous leishmaniasis in Brazil. Parasit Vectors 11:155

Dang LM, Piran J, Han D, Min K, Moon H (2019) A survey on internet of things and cloud computing for healthcare. Electronics 8:768

Dantas-Torres F, Brandão-Filho SP (2006) Visceral leishmaniasis in Brazil: revisiting paradigms of epidemiology and control. Rev Inst Med Trop Sao Paulo 48(3):151–156

Das ML (2004) Studies on Phlebotomus argentipes Annandale & Brunetti (Diptera: Psychodidae). Vector of kala-azar in Eastern Part of Nepal. PhD thesis submitted in BHU; Varanasi, pp 1–205

Das P, Samuels S, Desjeux P, Mittal A, Topno R, Siddiqui NA, Sur D, Pandey A, Sarnoff R (2010) Annual incidence of visceral leishmaniasis in an endemic area of Bihar, India. Tropical Med Int Health 15(Suppl 2):4–11

de Almeida AS, de Andrade MR, Werneck GL (2011) Identification of risk areas for visceral Leishmaniasis in Teresina, Piaui State, Brazil. Am J Trop Med Hyg 84(5):681–687

Deen M (1982) Geomorphology and land use: A case study of Mewat. Thesis submitted to the centre for the study of regional development, Jawaharlal Nehru University, New Delhi

Desjeux P (1996) Leishmaniasis. Public health aspects and control. Clin Dermatol 14:417–423

Desjeux P (2001) The increase risk for leishmaniasis worldwide. Trans Soc Trop Med Hyg 95:239–241

Desjeux P, Ramesh V (2011). Post-kala-azar Dermal Leishmaniasis: Facing the Challenge of Eliminating Kala-Azar from South Asia. Kala Azar in South Asia: Current Status and Challenges Ahead, edited by T.K. Jha, E. Noiri. page no. 111–120, DOI:10.1007/978-94-007-0277-6_11

Dhiman RC, Dinesh DS (1992) An experimental study to find out the source of fructose to sandflies. Indian J Parasitol 16:159–160

Dhiman RC, Shetty PS, Dhanda V (1983) Breeding habitats of phlebotomine sand flies in Bihar, India. Indian J Med Res 77:29–32

Dietrich D, Dekova R, Davy S, Fahrni G, Geissbühler A (2018) Applications of space technologies to global health: scoping review. J Med Internet Res 20(6):e230

Dinesh DS, Dhima RC (1991) Plant source of fructose to sandflies, particularly *Phlebotomus argentipes* in nature. J Commun Disord 23:160–161

Dinesh DS, Ranjan A, Palit A, Kishore K, Kar SK (2001) Seasonal and nocturnal landing/biting behaviour of *Phlebotomus argentipes* (Diptera: Psychodidae). Ann Trop Med Parasitol 95(2):197–202

Ding F, Wang Q, Fu J, Chen S, Hao M, Ma T, Zheng C, Jiang D (2019) Risk factors and predicted distribution of visceral leishmaniasis in the Xinjiang Uygur autonomous region, China, 2005–2015. Parasit Vectors 12:528

Duthie MS, Favila M, Hofmeyer KA et al (2016) Strategic evaluation of vaccine candidate antigens for the prevention of Visceral Leishmaniasis. Vaccine 34(25):2779–2786

Dye C (1996) The logic of visceral leishmaniasis control. Am J Trop Med Hyg 55:125–130

Dye C, Williams B (2010) The population dynamics and control of tuberculosis. Science 328(5980):856–861

El Hassan AM, Khalil EAG, Elamin WM, El Hassan LAM, Ahmed ME, Musa AM (2013) Misdiagnosis and mistreatment of post-Kala-Azar dermal leishmaniasis. Hindawi Publishing Corporation. Case Rep Med 2013:351579. https://doi.org/10.1155/2013/351579

El-Hassan AM, Ghalib HW, Zylstra E, Eltoum IA, Ali MS, Ahmed HMA (1990) Post-kala-azar dermal leishmaniasis in the absence of active visceral leishmaniasis. Lancet 336(8717):750

Elliot J (1863) Report on epidemic remittent and intermittent fever occurring in parts of Burdwan and Neddea divisions. Bengal Secretariat Office, Calcutta, India, pp. 1–23

Elnaiem DA (2011) Ecology and control of the sand fly vectors of Leishmania donovani in East Africa, with special emphasis on Phlebotomus orientalis. J Vector Ecol 36(1):S23–S31

Elnaiem DA, Connor SJ, Thomson MC, Hassan MM, Hassan HK, Aboud MA, Ashford RW (1998) Environmental determinants of the distribution of Phlebotomus orientalis in Sudan. Ann Trop Med Parasitol 92:877–887

Elnaiem DA, Hassan HK, Ward RD (2002) Associations of *Phlebotomus orientalis* and other sandflies with vegetation types in the eastern Sudan focus of kala-azar. Med Vet Entomol 13(2):198–203

Elnaiem DA, Schorscher J, Bendall A, Obsomer V, Osman ME, Mekkawi AM, Connor SJ, Ashford RW, Thomson MC (2003) Risk mapping of visceral leishmaniasis: the role of local variation in rainfall and altitude on the presence and incidence of kala-azar in eastern Sudan. Am J Trop Med Hyg 68(1):10–17

Feliciangeli MD, Delgado O, Suarez B, Bravo A (2006) Leishmania and sand flies: proximity to woodland as a risk factor for infection in a rural focus of visceral leishmaniasis in west central Venezuela. Tropical Med Int Health 11:1785–1791

Fernández MM, Malchiodi EL, Algranati ID (2011) Differential effects of paromomycin on ribosomes of Leishmania mexicana and mammalian cells. Antimicrob. Agents Chemother 55(1):86–93

Fitzpatrick C, Nwankwo U, Lenk E et al (2017) Chapter 17: An investment case for ending neglected tropical diseases. In: Holmes KK, Bertozzi S, Bloom BR et al (eds) Major infectious diseases, 3rd edn. The International Bank for Reconstruction and Development/The World Bank, Washington, D.C. https://doi.org/10.1596/978-1-4648-0524-0/ch17. Available from: https://www.ncbi.nlm.nih.gov/books/NBK525199/

Foley DH, Wilkerson RC, Dornak LL, Pecor DB, Nyari AS, Rueda LM, Long LS, Richardson JH (2012) SandflyMap: leveraging spatial data on sand fly vector distribution for disease risk assessments. Geospat Health 6(3):S25–S30

Fradelos EC, Papathanasiou IV, Mitsi D, Tsaras K, Kleisiaris CF, Kourkouta L (2014) Health based Geographic Information Systems (GIS) and their applications. Acta Inform Med 22(6):402–405. https://doi.org/10.5455/aim.2014.22.402-405

Fukuoka Y, Gay CL, Joiner KL, Vittinghoff EA (2015) Novel diabetes prevention intervention using a mobile app: a randomized controlled trial with overweight adults at risk. Am J Prev Med 49(2):223–237

Gahegan M (2000) Visualization as a tool for geocomputation. In: Openshaw S, Abrahart RJ (eds) Geocomputation. Taylor and Francis, London, pp 253–274

Galati EAB, Nunes VLB, Cristaldo G, Rocha HC (2003) Aspectos do comportamento da fauna flebotomínea (Diptera:Psychodidae) em foco de leishmaniose visceral e tegumentar na Serra da Bodoquena e área adjacente, Estado de Mato Grosso do Sul, Brasil. Rev Pat Trop 32:235–261

Garlapati RB, Abbasi I, Warburg A, Poché D, Poché R (2012) Identification of blood meals in wild caught blood fed Phlebotomus argentipes (Diptera: Psychodidae) using cytochrome b PCR and reverse line blotting in Bihar, India. J Med Entomol 49(3):515–521. https://doi. org/10.1603/ME11115

GBD (2013) Mortality and Causes of Death Collaborators. Global, regional, and national age-sex specific all-cause and cause-specific mortality for 240 causes of death, 1990–2013: a systematic analysis for the Global Burden of Disease Study 2013. *Lancet* 385(9963):117–171

Gebre-Michael T, Balkew M, Berhe N, Hailu A, Mekonnen Y (2010) Further studies on the phlebotomine sandflies of the kala-azar endemic lowlands of Humera-Metema (north-west Ethiopia) with observations on their natural blood meal sources, Parasit Vectors, 3, p. 6

Ghosh KN, Bhattacharya A (1989) Laboratory colonization of Phlebotomus argentipes (Diptera: Psychodidae). Insect Sci Appl 10:551–555

Gibson G, Torr SJ (1999) Visual and olfactory responses of haematophagous Diptera to host stimuli. Med Vet Entomol 13:2–23

Githeko AK, Lindsay SW, Confalonieri UE, Patz JA (2000) Climate change and vector-borne diseases: a regional analysis. Bull World Health Organ 78(9):1136–1147

Glynn LG, Hayes PS, Casey M, Glynn F, Alvarez-Iglesias A, Newell J, ÓLaighin G, Heaney D, O'Donnell M, Murphy AW (2014) Effectiveness of a smartphone application to promote physical activity in primary care: the SMART MOVE randomised controlled trial. Br J Gen Pract 64:e384–e391

Golpayegani AA, Moslem AR, Akhavan AA, Zeydabadi A, Mahvi AH, Allah-Abadi A (2018) Modeling of environmental factors affecting the prevalence of zoonotic and anthroponotic cutaneous, and zoonotic visceral leishmaniasis in foci of Iran: a remote sensing and GIS based study. J Arthropod-Borne Dis 12(1):41–66

Gonz'alez C, Wang O, Strutz SE, Gonz'alez-Salazar C, S'anchez-Cordero V, Sarkar S (2010) Climate change and risk of leishmaniasis in North America: predictions from ecological niche models of vector and reservoir species. PLoS Negl Trop Dis 4(1):e585

Guerin PJ, Olliaro P, Sundar S, Boelaert M, Croft SL, Desjeux P, Wasunna MK, Bryceson AD (2002) Visceral leishmaniasis: current status of control, diagnosis, and treatment, and a proposed research and development agenda. Lancet Infect Dis 2(8):494–501

Hailu A, Mudawi Musa A, Royce C, Wasunna M (2005) Visceral leishmaniasis: new health tools are needed. PLoS Med 2(7):e211. https://doi.org/10.1371/journal.pmed.0020211

Handler MZ, Patel PA, Kapila R, Al-Qubati Y, Schwartz RA (2015) Cutaneous and mucocutaneous leishmaniasis: clinical perspectives. J Am Acad Dermatol 73(6):897–908

Hartemink N et al (2011) Integrated mapping of establishment risk for emerging vector-borne infections: a case study of canine leishmaniasis in southwest France. PLoS One 6:e20817

Hertzman C (1994) The lifelong impact of childhood experience: a population health perspective. Daedalus 123(4):167–180

Jaffe CL, Rachamim N, Sarfstein R (1990) Characterization of two proteins from leishmania donovani and their use for vaccination against visceral leishmaniasis. J Immunol 144(2):699–706

Jervis S, Chapman LAC, Dwivedi S, Karthick M, Das A, Rutte EAL, Courtenay O, Medley GF, Banerjee I, Mahapatra T, Chaudhuri I, Srikantiah S, Hollingsworth TD (2017) Variations in visceral leishmaniasis burden, mortality and the pathway to care within Bihar, India. Parasit Vectors 10:601

Jeyaram A, Kesari S, Bajpai A, Bhunia GS, Krishna Murthy YVN (2012) Risk zone modelling and early warning system for visceral leishmaniasis (Kala-azar) disease in Bihar, India using remote sensing and GIS. The XXII congress of the International Society for Photogrammetry and Remote Sensing, 25 August–1st September, 2012, Melbourne Convection and Exhibition Centre

Joshi A, Narain JP, Prasittisuk C, Bhatia R, Hashim G, Jorge A, Banjara M, Kroeger A (2008) Can visceral leishmaniasis be eliminated from Asia? J Vector Borne Dis 45:105–111

Kalluri S, Gilruth P, Rogers D, Szczur M (2007) Surveillance of arthropod vector borne infectious diseases using remote sensing techniques: a review. PLoS Pathog 3(10):1361–1371

Kalra NL, Bang YN (1988) Manual of entomology in visceral leishmaniasis. World Health Organization, Document SEA/VBC/35. Regional Office for Southeast Asia, New Delhi

Kamilaris A, Ostermann FO (2018) Geospatial analysis and the internet of things. ISPRS Int J Geo-Inf 7:269. https://doi.org/10.3390/ijgi7070269

Kasap OE, Alten B (2006) Comparative demography of the sand fly Phlebotomus papatasi (Diptera: Psychodidae) at constant temperatures. J Vector Ecol 31:378–385

Kesari S, Kishore K, Palit A et al (2000) An entomological field evaluation of larval biology of sandfly in Kala-azar endemic focus of Bihar—exploration of larval control tool. Journal of Communicable Diseases 32, 284–288

Kesari S, Bhunia GS, Kumar V, Jeyaram A, Ranjan A, Das P (2010) Study of house-level risk factors associated in the transmission of Indian Kala-azar. Parasites & Vectors, 3:94

Kesari S, Bhunia GS, Kumar V, Jeyaram A, Ranjan A, Das P (2011) A comparative evaluation of endemic and non-endemic region of visceral leishmaniasis (Kala-azar) in India with ground survey and space technology. Mem Inst Oswaldo Cruz, Rio de Janeiro 106(5):515–523

Kesari S, Bhunia GS, Chatterjee N, Kumar V, Mandal R, Das P (2013) Appraisal of Phlebotomus argentipes habitat suitability using a remotely sensed index in the kalaazar endemic focus of Bihar, India. Mem Inst Oswaldo Cruz 108(2):197–204 doi:10.1590/0074-0276108022013012

Kienberger S, Hagenlocher M (2014) Spatial-explicit modeling of social vulnerability to malaria in East Africa. Int J Health Geogr 13:29

Killick-Kendrick R (1987) Investigation of Phlebotomine sand flies. In: Lumsden WHR, Evans DA (eds) Biology of the Kinetoplastids, vol 2. Academic Press, London

Krishnaswamy J, Bawa KS, Ganeshaiah KN, Kiran MC (2009) Quantifying and mapping biodiversity and ecosystem services: utility of a multi-season NDVI based Mahalanobis distance surrogate. Remote Sensing of Environment 113(4):857–867

Koch LK, Kochmann J, Klimpel S, Cunze S (2017) Modeling the climatic suitability of leishmaniasis vector species in Europe. Sci Rep 7:13325. https://doi.org/10.1038/s41598-017-13822-1

Kumar V, Kesari S, Sinha NK, Palit A, Ranjan A, Kishore K (1995) Field trial of an ecological approach for the control of Phlebotomus argentipes using mud and lime plaster. Indian J Med Res 101:154–156

Lainson R, Rangel EF (2005) Lutzomyia longipalpis and the eco-epidemiology of American visceral leishmaniasis, with particular reference to Brazil – a review. Mem Inst Oswaldo Cruz, Rio de Janeiro 100(8):811–827

Lane RP (1993) Sandflies (*Phlebotominae*). In: Lane RP, Crosskey RW (eds) Medical insects and arachnids. Chapman and Hall, London, pp 78–119

Lee M, Lee H, Kim Y, Kim J, Cho M, Jang J, Jang H (2018) Mobile app-based health promotion programs: a systematic review of the literature. Int J Environ Res Public Health 15(12):2838

Lengeler C (2004) Insecticide-treated bed nets and curtains for preventing malaria. Cochrane Database Syst Rev 2:CD000363

Lewis DJ (1971) Plebotomid Sandflies. Bull World Health Organ 44(4):535–551

Lewis DJ (1978) The phlebotomine sand flies (Diptera: Psychodidae) of the Oriental region. Bulletin of British Museum (Nature History), Entomology 37:217–343

Lima AP, Minelli L, Teodoro U, Comunello E (2002) Tegumentary leishmaniasis distribution by satellite remote sensing imagery, in Paraná State, Brazil. An Bras Dermatol 77(7):681–692

Lorenz C, Sergio A, Suesdek L (2015) Artificial neural network applied as a methodology of mosquito species identification. Acta Trop 152:165–169

Luan H, Law J (2014) Web GIS-based public health surveillance systems: a systematic review. ISPRS Int J Geo-Inf 2014(3):481–506. https://doi.org/10.3390/ijgi3020481

Luo EJ, Levitt L (2008) Massive splenomegaly. Hospital Physician, pp 31–38

Lysenko AJ (1971) Distribution of leishmaniasis in the Old World. Bull WHO 44:515–520

Mackie PF (1914) A flagellate infection of sandflies. Indian J Med Res 2:377–379

Madoff LC (2004) ProMED-mail: an early warning system for emerging diseases. Clin Infect Dis 39:227–232

Maia-Elkhoury ANS, Alves WA, Sousa-Gomes ML, Sena JM, Luna EA (2008) Visceral leishmaniasis in Brazil: trends and challenges. Cad Saúde Pública 24:2941–2947

Malaviya P (2015) Management of visceral leishmaniasis in Muzaffarpur, Bihar, India. PhD thesis, University of Antwerp

Mandal R, Das P, Kumar V, Kesari S (2017) Spatial distribution of *Phlebotomus argentipes* (Diptera: Psychodidae) in eastern India, a case study evaluating multi-spatial resolution remotely sensed environmental evidence and microclimatic data. J Med Entomol. https://doi.org/10.1093/jme/tjw232

Mandal R, Kesari S, Kumar V, Das P (2018) Trends in spatio-temporal dynamics of visceral leishmaniasis cases in a highly endemic focus of Bihar, India: an investigation based on GIS tools. Parasit Vectors 11:220

Mandal R, Kumar V, Kesari S, Das P (2019) Assessing the combined effects of household type and insecticide effectiveness for kala-azar vector control using indoor residual spraying: a case study from North Bihar, India. Parasit Vectors 12:409

Manson-Bahr PEC, Apted FIC (1982) Leishmaniasis. In: Manson-Bahr PEC, Apted FIC, eds. Manson's tropical diseases.18th ed. London: ELBS and Bailliere Tindall, 93–115

Maroli M, Feliciangeli MD, Bichaud L, Charrel RN, Gradoni L (2012) Phlebotomine sandflies and the spreading of leishmaniases and other diseases of public health concern. Med Vet Entomol 27:123–147

Marzochi MC, Fagundes A, Andrade MV, Souza MB, Madeira MF, Mouta-Confort E, Schubach AO, Marzochi KBF (2009) Visceral leishmaniasis in Rio de Janeiro, Brazil: eco-epidemiological aspects and control. Rev Soc Bras Med Trop 42:570–580

Meheus F, Boelaert M (2010) The burden of visceral leishmaniasis in South Asia. Trop Med Int Health 15(2):1–3

Meheus F, Boelaert M, Baltussen R, Sundar S (2006) Costs of patient management of visceral leishmaniasis in Muzaffarpur, Bihar, India. Trop Med Int Health 11:1715–1724

Mehraeen E, Ghazisaeedi M, Farzi J, Mirshekari S (2017) Security challenges in healthcare cloud computing: a systematic review. Global J Health Sci 9(3):157–166

Meinecke CK, Schottelius J, Oskam L, Fleischer B (1999) Congenital transmission of visceral leishmaniasis (kala azar) from an asymptomatic mother to her child. Pediatrics 104:e65

Mendes WS, Silva AA, Trovão JR, Silva AR, Costa JM (2002) Expansão espacial da leishmaniose visceral americana em São Luis, Maranhão, Brasil. Rev Soc Bras Med Trop 35:227–231

Merino-Espinosa G et al (2016) Differential ecological traits of two Phlebotomus sergenti mitochondrial lineages in southwestern Europe and their epidemiological implications. Trop Med Int Health 21:630–641

Miranda C, Massa JL, Marques CCA (1996) Analysis of the occurrence of American Cutaneous Leishmaniasis in Brazil by remote sensing satellite imagery. Rev Saude Publica 30(5):433–437

Morales MA, Cruz I, Rubio JM et al (2002) Relapses versus reinfections in patients coinfected with Leishmania infantum and human immunodeficiency virus type 1. J Infect Dis 185:1533–1537

Mott KE, Nuttall I, Desjeux P, Cattand P (1995) New geographical approaches to control of some parasitic zoonoses. Bull World Health Organ 73(2):247–257

Müller GC, Kravchenko VD, Rybalov L, Schlein Y (2011) Characteristics of resting and breeding habitats of adult sand flies in the Judean Desert. J Vector Ecol 36:195–205

Murray HW, Berman JD, Davies CR, Saravia NG (2005) Advances in leishmaniasis. Lancet 366:1567–1577

Nieto P, Malone JB, Bavia ME (2006) Ecological niche modeling for visceral leishmaniasis in the state of Bahia, Brazil, using genetic algorithm for rule-set prediction and growing degree day-water budget analysis. Geospat Health 1(1):115–126

Orshan L, Elbaz S, Ben-Ari Y, Akad F, Afik O, Ben-Avi I et al (2016) Distribution and dispersal of Phlebotomus papatasi (Diptera: Psychodidae) in a zoonotic cutaneous leishmaniasis focus, the Northern Negev, Israel. PLoS Negl Trop Dis 10:e0004819

Oryan A, Akbari M (2016) Worldwide risk factors in leishmaniasis. Asian Pac J Trop Med 9(10):925–932

Quinnell RJ, Courtenay O (2009) Transmission, reservoir hosts and control of zoonotic visceral leishmaniasis. Parasitology 136(14):1915–1934

Ouyang TH, Yang EC, Jiang JA, Lin TT (2015) Mosquito vector monitoring system based on optical wingbeat classification. Comput Electron Agric 118:47–55

Özbel Y, Sanjoba C, Matsumoto Y (2016) Geographical distribution and ecological aspect of sand fly species in Bangladesh. Kala Azar in South Asia 199–209

Parrot L (1936) Notes Le sur les Phlébotomes. XVII. Phlebotomus de 'Ethiopie. Arch Inst Paster d'Algérie 16:30

Parselia E, Kontoes C, Tsouni A, Hadjichristodoulou C, Kioutsioukis I, Magiorkinis G, Stilianakis NI (2019) Satellite earth observation data in epidemiological modeling of malaria, dengue and West Nile virus: a scoping review. Remote Sens 11:1862. https://doi.org/10.3390/rs11161862

Pascual Martı'nez F, Picado A, Roddy P, Palma P (2012) Low castes have poor access to visceral leishmaniasis treatment in Bihar, India. Trop Med Int Health 17(5):666–673

Patz JA, Graczyk T, Thaddeus K, Gellera N, Vittor AY (2000) Effects of environmental change on emerging parasitic diseases. Int J Parasitol 30:1395–1405

Petri G, Kennie TJM (1990) Terrain Modelling in Surveying and Civil Engineering. Whittles, London

Peterson AT, Pereira RS, Neves VFC (2004) Using epidemiological survey data to infer geographic distributions of leishmaniasis vector species. Rev Soc Bras Med Trop 37:10–14

Peterson AT, Soberón J, Pearson RG, Anderson RP, Martínez-Meyer E, Nakamura M et al (2011) Ecological niches and geographic distributions, Monographs in population biology, vol 49. Princeton University Press, Princeton

Picado A, Das ML, Kumar V, Dinesh DS, Rijal S, Singh SP, Das P, Coosemans M, Boelaert M, Davies C (2010a) *Phlebotomus argentipes* seasonal patterns in India and Nepal. J Med Entomol 47(2):283–286

Picado A, Das ML, Kumar V, Kesari S, Dinesh DS, Roy L, Rijal S, Das P, Rowland M, Sundar S, Coosemans M, Boelaert M, Davies CR (2010b) Effect of village-wide use of long-lasting insecticidal nets on visceral leishmaniasis vectors in India and Nepal: a cluster randomized trial. PLoS Negl Trop Dis 4(1):e587. https://doi.org/10.1371/journal.pnt-d.0000587

Picado A, Singh SP, Vanlerberghe V, Uranw S, Ostyn B, Kaur H et al (2011) Residual activity and integrity of PermaNet(®) 2.0 after 24 months of household use in a community randomized trial of long lasting insecticidal nets against visceral leishmaniasis in India and Nepal. Trans R Soc Trop Med Hyg 106(3):150–159

Picado A, Ostyn B, Singh SP, Uranw S, Hasker E, Rijal S et al (2014) Risk factors for visceral leishmaniasis and asymptomatic Leishmania donovani infection in India and Nepal. PLoS One 9:e87641

Picado A, Ostyn B, Rijal S, Sundar S, Singh SP, Chappuis F et al (2015) Long-lasting Insecticidal Nets to Prevent Visceral Leishmaniasis in the Indian Subcontinent; Methodological Lessons Learned from a Cluster Randomised Controlled Trial. PLoS Negl Trop Dis 9(4):e0003597. https://doi.org/10.1371/journal.pntd.0003597

Pigott DM, Golding N, Messina JP et al (2014) Global database of leishmaniasis occurrence locations, 1960–2012. Sci Data 1(1):140036

Poché D, Garlapati R, Ingenloff K, Remmers J, Poché R (2011) Bionomics of phlebotomine sand flies from three villages in Bihar, India. J Vector Ecol 36(S1):S106–S117

Poché DM, Grant WE, Wang HH (2016) Visceral Leishmaniasis on the Indian subcontinent: modelling the dynamic relationship between vector control schemes and vector life cycles. PLoS Negl Trop Dis 10(8):e0004868

Pollett S, Althouse BM, Forshey B, Rutherford GW, Jarman RG (2017) Internet-based biosurveillance methods for vector-borne diseases: are they novel public health tools or just novelties? PLoS Negl Trop Dis 11(11):e0005871

Pontryagin LS (1950) The mathematical theory of optimal processes and differential games. Trudy Matematicheskogo Instituta imeni VA Steklova 169:119–158, 1985

Prince SD (1999) What practical information about land-surface function can be determined by remote sensing? Where do we stand? In: Tenhunen LD, Kabat P (eds) Integrating hydrology, ecosystem dynamics, and biogeochemistry in complex landscapes. Wiley, Chichester, pp 39–60

Rab MA, Evans DA (1995) Leishmania Infantum in Himalayas. Trans Roy Soc Trop Med Hyg 89:27–32

Rahman SJ, Menon PK, Rajagopal R, Mathur KK (1986) Behaviour of Phlebotomus argentipes in the foothills of Nilgiris (Tamil Nadu), South India. J Commun Disord 18(1):35–44

Ranjan A, Sur D, Singh VP, Siddique NA, Manna B, Lal CS, Sinha PK, Kishore K, Bhattacharya SK (2005) Risk factors for Indian kala-azar. Am J Trop Med Hyg 73(1):74–78

Ready PD (2000) Sand fly evolution and its relationship to Leishmania transmission. Mem Inst Oswaldo Cruz 95(4):589–590

Ready PD (2010) Leishmaniasis emergence in Europe. Euro Surveill. 15(10):19505.

Ready PD (2013) Biology of phlebotomine sand flies as vectors of disease agents. Annu Rev Entomol 58:227–250

Rebelo JMM, Oliveira ST, Silva FS, Barros VLL, Costa JML (2001) Sandflies (Diptera: psychodidae) of the Amazonia of Maranhao v. seasonal occurrence in ancient colonization area and endemic for cutaneous leismaniasis. Rev Bras Biol 1(61):107–115

Reithinger R, Mohsen M, Leslie T (2010) Risk factors for anthroponotic cutaneous leishmaniasis at the household level in Kabul, Afghanistan. PLoS Negl Trop Dis 4(3):e639. https://doi.org/10.1371/journal.-pntd.0000639

Remaudière G (1992) A simplified method for mounting aphids and other small insects in Canada balsam. Rev Fr Entomol 14:185–186

Ribas LM, Zaher VL, Shimozako HJ, Massad E (2013) Estimating the Optimal Control of Zoonotic Visceral Leishmaniasis by the Use of a Mathematical Model. Article ID 810380, https://doi.org/10.1155/2013/810380

Rijal S, Uranw S, Chappuis F, Picado A, Khanal B, Paudel IS, Andersen EW, Meheus F, Ostyn B, Das ML, Davies C, Boelaert M (2010) Epidemiology of Leishmania donovani infection in high-transmission foci in Nepal. Tropical Med Int Health 15:21–28

Rogers L (1897) Report of an investigation of the epidemic of malarial fever in Assam or Kala-azar, Shillong, Assam. Secretariat Printing Office pp. 182–192

Rogers L (1950) Happy Toli: fifty-five years of tropical medicine, London, Muller. pp 29

Rogers DJ, Randolph SE, Snow RW, Hay SI (2002a) Satellite imagery in the study and forecast of malaria. Nature 415:710–715

Rogers ME, Chance ML, Bates PA (2002b) The role of promastigote secretory gel in the origin and transmission of the infective stage of Leishmania mexicana by the sandfly Lutzomyia longipalpis. Parasitology 124:495–508

Ross CE, Wu C (1995) The links between education and health. Am Sociol Rev 60:719–745

Rossi E, Rinaldi L, Musella V, Veneziano V, Carbone S, Gradoni L, Cringoli G, Maroli M (2007) Mapping the main Leishmania phlebotomine vector in the endemic focus of the Mt. Vesuvius in southern Italy. Geospatial Health 2: 191–198

Rotureau B, Joubert M, Clyti E, Djossou F, Carme B (2006) Leishmaniasis among gold miners, French Guiana. Emerg Infect Dis 12(7):1169–1170

Rouse JW, Haas RH, Schell JA, Deering DW, Harlan JC (1974) Monitoring the vernal advancements and retrogradation (greenwave effect) of nature vegetation. NASA/GSFC Final Report. NASA, Greenbelt

Rutledge LC, Ellenwood DA (1975) Production of Phlebotomine sandflies on the open forest floor in Panama: the species complement. Environ Entomol 4(1):471–477

Rutte EAL, Chapman LAC, Coffeng LE, Jervis S, Hasker EC, Dwivedi S, Karthick M, Das A, Mahapatra T, Chaudhuri I, Boelaert MC, Medley GF, Srikantiah S, Hollingsworth TD, de Vlasa SJ (2017) Elimination of visceral leishmaniasis in the Indian subcontinent: a comparison of predictions from three transmission models. Epidemics 18:67–80

Ryan JR, Mbui J, Rashid JR, Wasunna MK, Kirigi G, Magiri C, Kinoti D, Ngumbi PM, Martin SK, Odera SO, Hochberg LP, Bautista CT, Chan AS (2006) Spatial clustering and epidemiological aspects of visceral leishmaniasis in two endemic villages, Baringo District, Kenya. Am J Trop Med Hyg 74(2):308–317

Salomon OD, Orellano PW, Quintana MG, Pérez S, Sosa Estani S, Acardi S, Lamfri M (2006) Transmisión de la leishmaniasis tegumentaria en Argentina. Medicina (B Aires) 66:211–219

Sanyal RK (1985) Leishmaniasis in the Indian subcontinent. In: Chang KP, Bray RS (eds) Leishmaniasis. Elsevier Science. Publishers, Amsterdam, pp 443–467

Sanyal RK, Alam SN, Kaul SM, Wattal BL (1979a) Some observations on epidemiology of current outbreak of kala-azar in Bihar. J Commun Disord 11(4):170–182

Sanyal RK, Banerjee DP, Ghosh TK, Ghosh IN, Misra BS, Roy YP, Rao OK (1979b) A longitudinal review of kala-azar in Bihar. J Commun Disord 11(4):149–169

Schlein Y, Jacobson RL (1999) Sugar meals and longevity of the sand fly *Phlebotomus papatasi* in an arid focus of Leishmania major in the Jordan Valley. Med Vet Entomol 13(1):65–71

Schlerf M, Atzberger C, Hill J (2005) Remote sensing of forest biophysical variables using HyMap imaging spectrometer data. Remote Sens Environ 95:177–194

Scott HH (1939) A history of tropical medicine, London, Edward Arnold 2:1035

Seaman J, Mercer AJ, Sondorp E (1996) The epidemic of visceral leishmaniasis in Western Upper Nile, Southern Sudan: course and impact from 1984 to 1994. Int J Epidemiol 25:862–971

Sharma U, Singh S (2008) Insect vectors of Leishmania: distribution, physiology and their control. J Vector Borne Dis 45:255–272

Sharma AD, Bern C, Varghese B, Chowdhury R, Haque R, Ali M, Amann J, Ahluwalia IB, Wagatsuma Y, Breiman RF, Maguire JH, McFarland DA (2006) The economic impact of visceral leishmaniasis on households in Bangladesh. Trop Med Int Health 11:757–764

Sharma V, Purohit SK, Sharma G, Joshi R, Mehta RD, Kochar DK (2003) Cutaneous leishmaniasis in dogs and human beings in Bikaner, Rajashtan. Journal of Veterinary Public Health 1:69–73

Short HE, Barraud PJ, Craighead AC (1927) Studies on methods of transmission of kala-azar. Indian J Med Res 14(3):589–600

Shortt HE, Swaminath CS (1928) The method of feeding of *Phlebotomus argentipes* with relation to its bearing on the transmission of kala-azar. Indian J Med Res 15:827–836

Singh SP, Reddy DC, Rai M, Sundar S (2006) Serious underreporting of visceral Leishmaniasis through passive case reporting in Bihar, India. Trop Med Int Health 11(6):899–905

Singh A, Roy SP, Kumar R, Nath A (2008a) Temperature and humidity play a crucial role in the development of *P. argentipes*. J Ecophysiol Occup Health 8(1 & 2):47–52

Singh R, Lal S, Saxena VK (2008b) Breeding ecology of visceral leishmaniasis vector sandfly in Bihar state of India. Acta Trop 107:117–120

Sivagnaname N, Amalraj DD (1997) Breeding habitats of vector sand flies and their control in India. J Commun Disord 29(2):153–159

Smith G-H (1935) Relative relief of Ohio. Geog Rev 25:272–284

Snow J (1849) On the mode of communication of cholera. Churchill, London

Spickler AR, Roth JA, Galyon J, Lofstedt J (2010) Leishmaniasis (Cutaneous and Visceral): Edited by Emerging and Exotic Diseases of Animals). CFSPH Iowa State University, College of Veterinary Medicine, Ames, Lowa-50011, USA. ISBN: 978-0-9745525-8-3, pp. 205–210

Srividya A, Michael E, Palaniyandi M, Pani SP, Das PK (2002) A geostatistical analysis of the geographic distribution of lymphatic filariasis prevalence in southern India. American Journal of Tropical Medicine and Hygeine 67(6):480–489

Sudhakar S, Srinivas T, Palit A, Kar SK, Battacharya SK (2006) Mapping of risk prone areas of kala-azar (Visceral leishmaniasis) in parts of Bihar state, India: an RS and GIS approach. J Vect Borne Dis 43:115–122

Sundar S, Arora R, Singh SP, Boelaert M, Varghese B (2010) Household cost-of-illness of visceral leishmaniasis in Bihar, India. Trop Med Int Health 15(Suppl. 2):50–54

Tesh RB (1995) Control of zoonotic visceral leishmaniasis: is it time to change strategies? Am J Trop Med Hyg 52(3):287–292

Thakur CP (2000) Socio-economies of visceral leishmaniasis in Bihar (India). Trans R Soc Trop Med Hyg 94:156–157

Thakur CP, Thakur S, Narayan S, Sinha A (2008) Comparison of treatment regimens of kala-azar based on culture & sensitivity of amastigotes to sodium antimony gluconate. Indian J Med Res 127:582–588

Thompson RA, De Oliveira Lima JW, Maguire JH, Braud DH, Scholl DT (2002) Climatic and demographic determinants of American visceral leishmaniasis in northeastern Brazil using remote sensing technology for environmental categorization of rain and region influences on leishmaniasis. Am J Trop Med Hyg 67(6):648–655

Thompson RA, Maguire JH, de Oliveria Lima JW, Scholl DT, Braud DH (2004) Association of remotely sensed environmental indices with visceral leishmaniasis in Brazil. GISVET'04, 2nd International Conference on the Applications of GIS and Spatial Analysis to Veterinary Science, Held at the University of Guelph, Ontario Canada 23th-25th June 2004, p. 12–14.

Thomson MC, Elnaiem DA, Ashford RW, Connor SJ (1999) Towards a Kala-azar risk map for Sudan: mapping the potential distribution of *P. orientalis* using digital data of environmental variables. Trop Med Int Health 4(2):105–113

Thornton SJ, Wasan KM, Piecuch A, Lynd LLD, Wasan EK (2010) Barriers to treatment for visceral leishmaniasis in hyperendemic areas: India, Bangladesh, Nepal, Brazil and Sudan. Drug Dev Ind Pharm 36:1312–1319

Tobler WR (1970) A computer movie simulating urban growth in the Detroit region. Econ Geogr 46:234–240

Torres-Guerrero E, Quintanilla-Cedillo MR, Ruiz-Esmenjaud J, Arenas R (2017) Leishmaniasis: a review. F1000Res 6:750. Published 2017 May 26. https://doi.org/10.12688/f1000research.11120.1

Tsegaw T, Gadisa E, Seid A, Abera A, Teshome A, et al (2013) Identification of environmental parameters and risk mapping of visceral leishmaniasis in Ethiopia by using geographical information systems and a statistical approach. Geospat Health 7: 299–308

Turner-McGrievy GM, Beets MW, Moore JB, Kaczynski AT, Barr-Anderson DJ, Tate DF (2013) Comparison of traditional versus mobile app self-monitoring of physical activity and dietary intake among overweight adults participating in an mHealth weight loss program. J Am Med Inform Assoc 20:513–518

Twining W (1832) Clinical illustrations of the more important diseases of Bengal, with the result of an inquiry into their pathology and treatment, Calcutta. Baptist Mission Press, pp. 271–360

Viergever R, Perehudoff K, Esselink M, Sienkiewicz D, Panday BM (2005) Leishmaniasis: a neglected disease. The Dutch secretary of defense reports to parliament. Available at: http://www.globalmedicine.nl/index.php/leishmaniasis

Waitz Y, Paz S, Meir D, Malkinson D (2019) Effects of land use type, spatial patterns and host presence on Leishmania tropica vectors activity. Parasit Vectors 12:320. https://doi.org/10.1186/s13071-019-3562-0

Wang Q, Adiku S, Tenhunen J, Granier A (2005) On the relationship of NDVI with leaf area index in a deciduous forest site. Remote Sens Environ 94(2):465–474

Werneck, GL (2014) Visceral leishmaniasis in Brazil: rationale and concerns related to reservoir control. Revista de Saúde Pública 48(5):851–856

Werneck GL, Maguire JH (2002) Spatial modeling using mixed models: an ecologic study of visceral leishmaniasis in Teresina, Piauí State, Brazil. Cad Saude Publica 18(3):633–637

Williams P (1993) Relationships of Phlebotomine sand flies (Diptera). Mem Inst Oswaldo Cruz 88:177–183

World Health organization (WHO) (1991) What you need to know about vector borne disease. Available at: http://whqlibdoc.who.int/hq/1991/WHO_CWS_91.3_3_eng.pdf. Chapter II, pp. 5

World Health Organization (2004) Report of the scientific working group meeting on leishmaniasis. World Health Organization, Geneva

World Health Organization (2016) End in sight accelerating the end of HIV, Tuberculosis, Malaria and Neglected Tropical diseases in the southeast Asia region. World Health Organization, Regional Office for South-East Asia, Indraprastha Estate, Mahatma Gandhi Marg, New Delhi, India. ISBN 978-92-9022-527-0

World Health Organization (WHO) (2000a) The leishmaniasis and Leishmania/HIV co-infections. Fact sheet number 116. Geneva: World Health Organization

World Health Organization (WHO) (2000b) The leishmaniasis and Leishmania/HIV co-infections. Fact sheet number 116. Geneva: World Health Organization

World Health Organization (WHO) (2001) WHO recommended strategies for the prevention and control of communicable diseases. Department of Communicable Disease Control, Prevention and Eradication. WHO/CDS/CPE/SMT/2001.13. Available at: http://whqlibdoc.who.int/hq/2001 /WHO_CDS_CPE_SMT_2001.13.pdf

World Health Organization (WHO) (2002) Urbanization: an increasing risk factor for leishmaniasis. Wkly Epidemiol Rec 77:365–372

World Health Organization (WHO) (2010) Control of the leishmaniasis. Report of a meeting of the WHO Expert Committee on the control of Leishmaniasis, Geneva, 22–26 March, 2010. Available at: http://whqlibdoc.who.int/trs/WHO_TRS_949_eng.pdf

Zaman K, Yunus M, Arifeen SE et al (2006) Prevalence of sputum smear-positive tuberculosis in a rural area in Bangladesh. Epidemiol Infect 134(5):1052–1059

Zhang Z, Ward M, Gao J, Wang Z, Yao B, Zhang T, Jiang Q (2013) Remote sensing and disease control in China: past, present and future. Parasit Vectors 6:11

Zhao S, Kuang Y, Wu CH, Ben-Arieh D, Ramalho-Ortigao M, Bi K (2016) Zoonotic visceral leishmaniasis transmission: modeling, backward bifurcation, and optimal control. J Math Biol 73(6–7):1525–1560

Index